HOMO SAPIENS TO HOMO 'X'

WHY OUR CHILDREN AND YOUTH ARE DIFFERENT FROM US

A philosophical enquiry into the evolution
of man since archaic times

LAWRENCE NYAGUTI OCHIENG

PARTRIDGE
A Penguin Random House Company

To order additional copies of this book, contact
Toll Free 0800 990 914 (South Africa)
+44 20 3014 3997 (outside South Africa)
orders.africa@partridgepublishing.com

www.partridgepublishing.com/africa

Contents

Overview

The book, *Homo sapiens to Homo x* can be summed up into four parts, all connected to demonstrate that the changes in our children and youth today are not mere generational but evolutionary changes whose result is a new human species *Homo x*. The introductory pages which include the abstract and Introduction chapter of the book offers the reader theoretical background and developments in the evolution research relevant to prove this evolution thought.

Chapter 2 gives justifications leading to a conclusive proposition that a complete evolution metamorphosis of human beings has been able to occur within the last 150,000 years, a departure from previous evolution epochs that occurred or were noticed within the *law of fossil succession* to have occurred in Millions of years by correlating periods of evolution and intensity of what is defined as transient triggers.

Chapter 3, 4 and 5 explains *coexistence for survival* as a social theme that has facilitated survival against biological advantageous characteristics by demonstrating that for human being, survival has been more in collectively nurturing weakness or a mixture of weakness and advantageous traits rather than sole survival of the most advantaged groups or individuals, making mutualism the basis for continued survival more significant as opposed to biological characteristics inherent in the individuals in a group. These chapters offers the reader the historic dynamics that resulted to the great migration of the other groups from Africa and effect of the migration on the groups that settled out of Africa, then demonstrates how the re-unification of Human beings has hastened the evolution process, further demonstrating the succession environment that has fast tracked human beings to be able to survive and to device how to survive better thereby ushering in the next level hominid. It is not surprising that events such as wars, colonization, and slavery among others are assessed as impetus to human to human mutualism on the basis of the end results. Redistribution of roles by gender is further assessed on the basis of its evolutionary importance.

Finally Chapter 6 and 7 explains psychological, physiological, and socio-cultural changes that clearly distinguish the difference between our youth and children from us using illustrations and observatory narratives, thus, drawing a justification for re-classification of these groups as a distinct higher hominid or species. It is further projected that the re-unification of humans through globalization will biologically result into genetical combination of various races resulting into a hominid without racial distinction within the maturity period of this new hominid (Homo x) using *a genetical chart* demonstration.

Foreword

It was so enriching to meet Lawrence Nyaguti Ochieng the Author of *Homo sapiens to Homo x* at the University of Maastricht-Netherlands during my Masters in Public Policy and Human Development. Back then, the topic of the "welfare state" was in the centre of the discussion. For Lawrence and I, it was clear not only that immediate action is *necessary* to tackle the world's levels of poverty, inequality and social injustice, but also *unavoidable* as a prerequisite for the human evolution.

On this book, Lawrence analyses the fundamental causes of the human evolution, taking forward and giving a pragmatic intellectual understanding to Charles Darwins *Organic theory of evolution* to explain the state of Humans today and in future, with a more clear perspective that differentiates human beings from Darwinian descriptions of evolution on the basis of *Survival of the fittest*. He insists and provides proof that the entire human beings survival is dependent on support from the strong and the weak and vice versa.

Depending on lenses of the discipline you wear, you will find this book to the point. It is multidisciplinary. The most exciting part is the demonstration that humans have evolved using physiological, psychological and social behavioural evidence.

In a very oneiric and comprehensive manner, Lawrence brilliantly manages to put all these pieces into place. With a profound academic understanding of the topic he delivers a piece of work that is both nice to read and very instructive. I hope the reader enjoys it as much as I did.

Augusto CAOA-GOUDAILLIEZ- Argentina

Acknowledgements

My first acknowledgements are to my late father, E. A. Ochieng-Obado, for the inspiring interdisciplinary intellectual debates over the years that shaped and broadened my views so far. My late mother, Sarah Opado, to whom I owe my writing skills and the good English writing foundation.

My brothers—Edwin Obado, Hamisi Nyatieng, Samuel Osee (the forester), and Lazarus Monye—who always pushed me to start and finish this book project.

I also thank my kids—Michael, Cornel, Lawi, Sarah, and Jamal—for giving me some insights that have formed the basis of this book.

It was largely my wife, Jill Navalayo, who, besides moderating my anxiety, was able to give me peace of mind to complete this book.

For the consultants and friends who supported this aspiration, thank you, all!

Abstract

Homo sapiens to Homo x is social enquiry on the continuing evolution of the modern man, giving rise to a new human species (*Homo x*), who is the successor of *Homo sapiens sapiens*. It seeks to explain why the current generation of children and youths exhibit differences in behavioural, social, and (to some extent) physiological characteristics from us by exploiting Charles Darwin's organic theory of evolution and further differentiating the evolution of man as a unique occurrence from that of other animals and plants.

The evolution epoch of the *Homo sapiens*, which has undergone a full *hominid evolution metamorphosis*, has resulted to distinguished characteristics now noticeable in our children and youths. As opposed to survival on the basis of advantageous characteristics of individuals or groups of the archaic times, the progression of human evolution has been largely reliant on *coexistence for survival*

It is thus within the context of social changes across the earth and historical development over the years as recorded from the beginning of records by humans in addition to other biological facts of evolution largely consistent with the *organic theory of evolution* that I advance this hypothesis on evolution to prove the existence of a product of a complete evolution phase in the hominin evolution cycle. The results of this natural process makes our youth and children a new human species- Homo x.

CHAPTER 1

INTRODUCTION

The reason behind physiological differences between *Homo habilis* to *Homo sapiens* is best explained by the *organic theory of evolution*, which is the most acceptable theory of evolution for advancement of postulations on progressive changes in humans. Evidently, an affirmation to the organic theory of evolution has been proved by excavation of historical sites and discoveries of prehistoric and ancient fossils that obey the *laws of fossil succession*. Thus since the *Origin of Species* by Charles Darwin, fossil discoveries of *Australopithecus afarensis* in the 1980s in Ethiopia, australopithecines (the early hominids) and *Homo erectus* in the 1960s and 1970s in Turkana among other discoveries only served to prove the validity of the theory of gradual change in man and that man could have developed from an apelike ancestor.

The overarching principle is that change is constant and gradual. It is thus logical to assess the extent of this infinite process of change and delineate its progress and results at each stage. Therefore, within the compromise in acceptance that the *hominin* has been evolving over millions of years and that this process is ongoing, we seek to justify the discovery of a new human species. In the meantime, the species has been assigned the reference *Homo x* pending further consensus on the best <u>nomenclature</u> that defines the future humanity. Certainly, we are able to demonstrate that the emerging human species' social characteristics differ prominently from its predecessor, the *Homo sapiens sapiens* (generally referred here as the *Homo sapiens*). Further, the evolution time frame from a species to its next level has been delineated into stages that coincide with the progression of human history. The entire time frame is assigned the term *transiential period*[1] while the causes of evolution have been referred to as transiential triggers.

In other words, there is a clear departure in all social and cultural sense between the contemporary man; more visibly in our youth and children, the modern man, and the archaic man, clearly indicating the possibility that humans have changed. Whereas this study does not have a biological laboratory to authenticate the possibility of any biological changes occasioned or that triggers these social changes or vice versa, through social enquiry, we all admit that indeed there are phenomenon differences. Perhaps this is a stimulus for further scientific research to falsify or validate this *neo-Darwinist thought*.

This neosocial Darwinist thought is therefore diferent in approach and a departure from early social Darwinists of the late nineteenth and early twentieth centuries (Herbert Spencer 1960, Thomas Malthus 1798, Francis Galton 1965) who used the scientific theory (in natural selection) to justify their arguments that individuals and groups are subject to the same Darwinian laws of natural selection as plants

---

[1]   *Transiential period* is the period within which human beings transform their biological and sociocultural characteristics (discussed later).

and animals, resulting to the rationalization of political conservatism, imperialism, racism, and discouragement of intervention and reforms that would improve mutual relationship, cohesion, and unity of purpose for the survival of all humans both economically and socially, without making an evolutionary analysis and seeking for historical background behind various human distinctions..

It is on this premise that this book seeks to advance the Darwinian discovery to make a contemporary sense on the organic theory of evolution by utilizing existing social-cultural, historic and scientific dimensions to the evolution discourse. The result is to arrive at a perspective for not only understanding the infinite character of the evolution process but also to demonstrate the relationship between social and cultural behaviour in human beings since transition to Homo sapiens that have now produced a tangible result in our youth and children. Certainly, the results evident in assessment of human psychological, physiological and socio-cultural changes as demonstrated in Chapter 6 and 7. With archaeological facts and empirical data from carbon dating, scientific innovations, and historical narratives, it is now possible to track the historical, social, and cultural behaviour of *Homo sapiens*, which has now existed for approximately over 100,000 years, and thus make conclusive observation of the existing changes and forward projections on the future outcomes of these changes.

Human Origin and Evolution

While various religious and cultural myths of creation have ended at explaining how humans were created, it has not been possible to rationalize these myths to explain the basis for the biological, social, and cultural changes that have occurred to mankind progressively over the years.

According to a Chinese mythology, Nuwa began creating men from yellow clay, then dipped a rope in clay and flicked it, resulting to blobs of clay landing everywhere. And then each of these blobs became a person. The myth further seeks to explain social class differences by stating that nobles were created from the handcrafted figurines while commoners were created from the blobs. According to other cultural myths in Africa, a metaphysical being lowered the first pair or two of mankind from the clouds to the earth. They brought with them cattle, sheep, and goats, and the two pairs reproduced so that their children intermarried and formed families of mankind on earth (Kamba). Similarly, it is held among the Masai and Nandi of Kenya that men came originally from a leg or knee. This knee or leg belonged to some other being, evidently like men. The leg got swollen until finally it burst, letting out a male person on one side and a female on the other side.

The most popular myth of origin among Christians, Muslims, and Jews is the story of Adam and Eve. God or Allah created Adam in his own image and then felt that Adam was lonely and, as a result, created Eve in Adam's likeness but with a different physiology. The initial intention of creation, according to the Creation mythology, was that the two were intended to take care of the *Garden of Eden*. Reproduction in this popular myth therefore becomes a consequence of punishment. It should however be noted that in this popular myth, there is a step-by-step or progressive creation of the universe, and eventually man is created on the sixth day.

Thus the common thread in all creation myths is that creation was a process. Similarly, there are other scientific, philosophical, or metaphysical theories of the beginning of life, such as *cosmogony* (which is concerned with the coming into being of the universe), *abiogenesis* (life is formed from amino acids, which are organized into protein), *panspermia* (that life exists throughout the universe and is ejected from space by *meteors* and becomes active depending on conditions met), *symbiogenesis* (that different bacteria forms became symbiotic and formed eukaryotic cells). And others such as the materialistic theory, clay theory, consecutive creation theory, and spontaneous generation theories end at hypothesizing how life forms began on earth without also resolving the diversity of life forms to date and, most specifically, the changes in humans' biological, social, and cultural forms to date.

On the other hand, with the organic evolution theory, it is possible to explain the gradual changes over time in hereditary traits of man and thus the changing social cultural behaviour, which forms the basis of this theory.

So ever since Charles Robert Darwin formulated the organic theory of evolution (*Origin of Species, The Descent of Man*) that explains the divergence of life forms and gradual biological differences in the physiological characteristics of human beings, biological scientists, such Gregor Mendel, concluded that traits are heritable in a predictable manner. Thomas Henry Huxley demonstrated (in 1862 at a Cambridge meeting with William Flower) the relationship between humans and apes among others, and many archaeologists have been excavating the ground to come up with more concrete justifications, if not invalidations, for the dominance of the *evolution theory* in understanding the changes that have occurred over the years in both plants and animals, all ending up with the affirmation to the fact that evolution is indeed a continual process.

Whereas the organic theory of evolution is the best that can be utilized to arrive at the *Homo x* part, Darwin's explanations run on the basis of biological characteristics assumed to be advantageous depending on inheritable traits in a world view that is predominantly prejudiced by his European standpoint. His viewpoint on sociocultural issues of evolution presented in *The Descent of Man and*

Selection in Relation to Sex (1871) does not account for possible convergence of human sociocultural behaviour and integration, whose results relate to all humans, developed or not developed and either civilized or uncivilized. It could have further misled to the conclusion that the 'most civilized is the fittest to survive' on an assumption that the racial variance in civilization then was an indicator of the most advantageous of humans, resulting to the racial and anti-reformist conclusions of early social Darwinists.

Be that as it may, this thought is developed from an outstanding convergence of the fact that Darwin noted:

> Although the existing races of man differ in many respects, as in colour, hair, shape of skull, proportions of the body, &c., yet if their whole organisation be taken into consideration they are found to resemble each other closely in a multitude of points. Many of these points are of so unimportant or of so singular a nature that it is extremely improbable that they should have been independently acquired by aboriginally distinct species or races. The same remark holds good with equal or greater force with respect to the numerous points of mental similarity between the most distinct races of man. The American aborigines, Negroes and Europeans differ as much from each other in mind as any three races that can be named; yet I was incessantly struck, whilst living with the Fuegians on board the Beagle, with the many little traits of character, shewing how similar their minds were to ours; and so it was with

a full-blooded negro with whom I happened once to be intimate.[2]

We can now link the thread of the whole hominin evolution to a collective social and cultural behaviour in space and time thanks to globalization that has caused a relatively similar *universal characteristic* of all races of the world. At least, there are common sociocultural and behavioural expectations of each human by all regardless of the environment, culture, and race. Furthermore, it is now anticipated that all *men are decent*, and thus observations reveal that social and cultural behavioural changes of these decent men could be a pointer to major or minor biological and psychological changes occurring, which has affected cultural mannerisms and communal behaviour.

Development in Evolution Research

Whereas Charles Darwin's recorded observations and hypothesis was out of his experiences during the five-year *survey voyage* of the HMS *Beagle* (1831–1836), over the last fifty years, archaeologists and other scientific expeditions have benefited from technological advancements occasioned by innovation of more efficient archaeological research methods, such as remote sensing and virtual archaeology among others and tools such as drones, which have sped up archaeological and scientific enquiries beneath the grounds of targeted human historical sites to rediscover that path to the modern man, *Homo sapiens sapiens.* The world view for researchers has further been broadened by opening up different regions of the world through infrastructural development, security, and reduced communication barriers. During the same period, there has been increased social

---

[2]   Darwin 1871, pp. 231–232, Vol. 1.

incentives and rewards associated with discoveries of fossils of prehistoric creatures and the resultant postulations. Most of these discoveries, however, are proof that the Darwinian wisdom has subdued the modern archaeologists and natural science to confinement within the sense of the popular (*Darwin's*) evolution theory.

The development of the social interpretation to Charles Darwin's evolution theory is majorly traced to Herbert Spencer, whose interpretation (*natural selection* and *survival of the fittest*) in application to society asserts that humanitarian impulses had to be resisted as nothing should be allowed to interfere with nature's laws, including the social struggle for existence (*Principles of Biology* (1864)). The social Darwinist exploited the world view for which Darwin's writings were staged and his unsubstantiated argument on what he describes as the 'civilized man'.[3]

So far, taking a social scientific dimension, this *neosocial Darwinist thought* exploits social features existing in every successive human species to arrive at *the new biological discovery* that is a departure from the distortion of *social Darwinism* of the early to mid nineteenth century.

With hindsight to the natural selection theory, early social scientists, taking some cue from *Darwin's* lack of clarity in the use of social terms such as *civilized man*, misinterpreted the natural selection theory to explain cultural variance. This earlier mis/interpretation of *survival of the fittest* led to the classification of the world into civilized and uncivilized, justifying colonization of states, wars, and economic savagery. In this case, the uncivilized were assumed to represent the disadvantaged, who became victims of social injustices and were therefore justified to become extinct.

---

[3]   As used in *Origin of Species* to describe some humans and also in the *Decent of Man Vol. 1* (Darwin 1871, pp. 200–201).

However, critics were unable to give a logical rejoinder to these social dimensions offered by the *social Darwinists* within the evolution context but instead made their (of social Darwinists) discourse socially unfashionable—without counter—factual arguments.

In this book, the social Darwinists' logic has now been advanced within the *neosocial Darwinist* thought to a higher and objective argument to explain the difference in socio political and cultural changes in Africa, Europe, Asia, and the American continents during the theoretical environment within which Charles Darwin and other writers lived up to the twenty-first century. This approach has integrated both scientific and social changes recorded over time to demonstrate that we have actually evolved.

CHAPTER 2

THE EVOLUTION PROCESS OF SAPIENS

Intensity of Lifespan Experience versus Duration of Evolution in Hominins

Lifespan experience is the intensity of interaction in the environment within the length of time that a hominid lives. Duration of evolution is a given time frame within which evolution occurs. Life expectancy falls within the duration and is based on the individual's period between birth and death. Lifespan experience is inversely proportional to the duration of evolution. The more intensive the lifespan experience, the lesser the duration of evolution of a hominin.

Intensity implies a variety of environmental experiences that would either alter or advance the course of human survival by either facilitating progressive survival or threatening extinction. These environmental experiences are the triggers of evolution, such as climatic changes, technology development, human interaction, and diseases among others, which then increase the chances for adaptation. Thus each successive hominid has had different lifespan experiences due to the variation of experiences in geographical and historical times they lived. On the same note, hominids in similar historical times but different environments have had different lifespan experiences, as explained in further discussions.

Homo habilis, the *handy man*, existed between 2.4 and 1.5 million years ago whereas *Homo erectus* lived between 1.8 million and 300,000 years ago implying that *Homo—erectus* evolution period to the next level was relatively longer compared to its predecessor *Homo sapiens* (archaic) provides the bridge between *Homo erectus* and *Homo sapiens sapiens* during the period 500,000 to 200,000 years ago, *Homo sapiens sapiens* 120,000 years ago. It is noted that based on this theory, Homo Sapiens Sapiens arises from recognition that there was phenomenalpace of changes experienced in the life of Homo Sapiens.

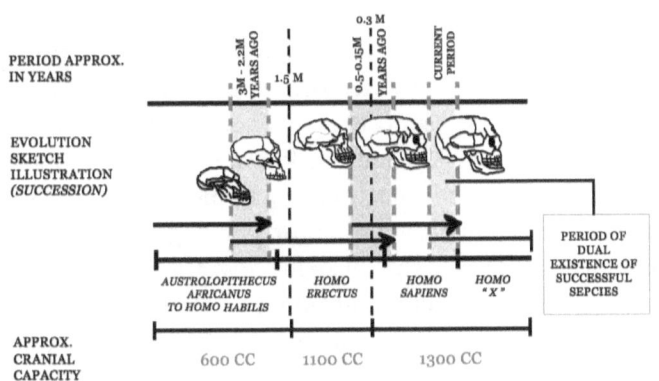

Evolution graph.[4]

The duration of development in the evolution from one level of the species to another is different depending on the historical period. It is evident therefore from the evolution

---

[4]   http://www.serpentfd.org/section2hominidevolution.html.

graph that the *Homo erectus* had the highest duration in its evolution process. In other words, the *H. erectus* had a less intensive lifespan experience arising from exposure to a less threatening environment or staying in a routine state with fewer triggers of evolution, thereby delaying its evolution process while transforming marginally with time.

Probably the early *Homo sapiens* (archaic) was a continuation of *Homo erectus* at its maturity stage of evolution. The climatic and environmental conditions must have been favourable for his survival. He had tools that possibly moved him to a life comfort zone that did not trigger meaningful species development.

The period of dual existence by *Homo erectus* and *Homo sapiens* (archaic) was quicker than *Homo habilis* to *Homo erectus*. Tracing the period historically, human to human interactions had begun making man adapt for survival thus ushering in the *Homo sapiens sapiens*, or the modern *Homo sapiens*. Today, the dynamic state of nature has further shortened the lifespan experience of the human species. Humans' exposure to experiences like climate changes, interaction, conflicts, knowledge, and technology happens within a shorter time due to increased population, change in social behaviour, and technological innovations. It is most likely that the cause for migration of the early or archaic *sapiens* was human-to-human conflicts resulting to displacements and experiences that enabled the migrating groups to adapt resilience to overcome harsh environmental challenges, creating racial differences.

To understand lifespan experience, take an average person in the year AD 1800 in an environment of little technological advancement in mobility and communication as well as relatively low human exposure and understanding of the environment, then compare that person to someone living in AD 2014, which is characterized by more human interactions, advanced technological development, accessible geographical environment, and liberal cultural practices. You will realize that it can now take human beings only an average of one day or less to get exposure or environmental experience that people living in AD 1800 could have in a year.

Or if you are a parent today, you will realize that your kids are able to relate with most modern technology, like mobile phones, computers, *et cetera*, quicker than you could. They are exposed to television programs even those of adult content however we try to restrain them. From age 5 of their lives, they are at par with us in understanding basic technology and global exposure. They follow up on soap operas and news and are able to relate to various issues as presented directly or as implied and are updated on persistent changes daily.

Lifetime exposure of the contemporary man is higher, and only one year of exposure today could have the same intensity of exposure comparable to thousands of years for the *Homo habilis* or *Homo erectus*. This supports the conclusion that while it has been historically and scientifically proven that it took millions and thousands of years for the evolution process on man to take place, there is a higher possibility or certainty that the increased lifespan experience of the species drastically reduced the evolution period of the modern *sapiens*.

The *contemporary man* seems to exhibit radical differences in interacting with nature in terms of *habitat, tools, knowledge, language, survival culture,* and *social instincts* than the modern man. This means that the early modern man is on his transition period, and its duration depends on how he prepares a succession ecosystem for his successor. Indeed, there are signs that the contemporary man has begun to relinquish his dominance to allow for the survival of the next human species in an environment that he is currently dominating. It is upon this understanding that it can now be concluded that there is a definite possibility for *Homo sapiens* (modern human) to have reached its *intra-transient* maturity stage, therefore managing the human species succession to *Homo x*.

Human Transiential Period

The human transiential period is the period within which human beings transform their biological and sociocultural characteristics. For the purpose of understanding this concept, it is illustrated that there are two transient levels in the evolution of the human species—that is, *intratransiential* and *intertransiential periods*. While the former represents the period that accounts for developments that do not necessarily result into a new species, the latter is the period of successive coexistence of a species and its progenitor.

To further explain the transiential period, it is observed that there are three stages of the *intratransiential stage*— namely, *archaic stage*, *rational consciousness stage*, and the *maturity stage*.

The *archaic stage* is the first level of transformation, where the new species emerges. This stage is the most critical because there is most likely coexistence of a species with its progenitor in transition; in this case, the progenitor would have transcended from the *maturity stage* of its species development. For example, the dates of the archaeological discovery of *Homo erectus* indicate that they could have coexisted with *Homo sapiens* for a span of time before the complete disappearance of the former. The archaic stage is mostly characterized by lack of self-consciousness and poor memorization of the surrounding environment. At the first stage of transition to the next level, humans are solitary, mechanical, genuine, uncertain, and therefore driven by fear of the unknown. The mode of survival in this stage for the *Homo sapiens* was mostly as hunter-gatherers, and kinship institutional behaviour is similar to other animals' territorial behaviour. At this stage, there is no fixed habitat as man is majorly nomadic.

The *rational consciousness stage* is the stage for self-discovery, universal conceptualization, and settlement. This stage begins the institutionalization of communalism, cultural development, settlement, and territorialism; mystification led to the development of religion, the advancement of technology and scientific perspectives, and finally, the mastery of cosmology. This stage is characterized by cultural development and identity search, scientific innovations, territorial expansion, domestication of animals, intercultural interactions, and humanistic movements. As a consequence of rationalization, parasitic and symbiotic social behaviours are exhibited at this stage. It is suggested here that there is a possibility that these characteristics, which could have begun some 50,000 years ago to date, sum up the archaeological shift from *Homo sapiens* to *Homo sapiens sapiens*, and here we consider this part of the rational consciousness stage of the *Homo sapiens*.

The final stage is the *maturity stage*, which begins with globalization, advancement of technology, and virtual conceptualism. The maturity stage is the peak of internationalized human consciousness and is the stage that prepares for a succession ecosystem to usher a new species. This is when humans prepare for the intertransiential stage. The transformation from archaic stage to rational consciousness stage and then to maturity stage represents the intertransiential stage.

The transformational history of the *Homo sapiens* can thus be illustrated within the *transiential hypothesis* to approximate classified historical periods beginning from *prehistoric ages* and running to *ancient times, post-classical times,* and *modern times* and through to *contemporary times*. It should furthermore be noted that each historical epoch representing a stage within the *transiential* period is most likely characterized by the development level bequeathed by the predecessor during the *intertransiential stage*. For example, the current level of development of humans dictates the standards and historical characteristics of the next generation, which is supposed to be more advanced if not more adaptive to the environment.

Although it is a challenge to demonstrate the different stages of development of *H. habilis* and *H. erectus* within the *transient cycle* hypothesis due to the lack of social definition of the historical periods and lack of recorded socio-cultural history of the periods of their existence, the *Homo sapiens* offers the theory credence to its periodic growth experienced between the *archaic* and *maturity* stages of its transient cycle. It is, however, left for further scientific and anthropological *excavations* to find fossils that justify this theory to complete the puzzle of the development of the earlier human species.

Undoubtedly, *H. sapiens* have drastically transformed physically, socially and culturally during the *intratransiential* stages to enable further conclusions based on the usual drastic differences of evolving species. Further, the historical periods equated within the *transiential theory* is such that the *archaic to medieval historical periods* represent the *archaic transiential stage*, the *rational consciousness transiential stage* represents the period from the *post-medieval* up to the *modern historical times*, and the *maturity transiential stage* represents from the period from the *contemporary* to the *future historical period*. Therefore, to study the transformation of *Homo sapiens* is to study the transformation of human beings in the historical times through the lenses of evolution.

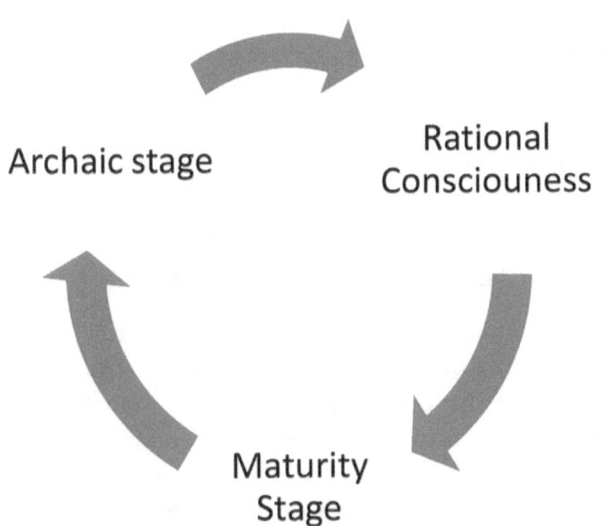

Intratransient evolution cycle.

Transiential Triggers

Transiential triggers are the catalysts for development of the human species from one stage of the *transiential period* to another. Evolution of mankind to date and to the next level can be attributed to stimulation by these triggers. The major stimuli are environmental factors, threat by diseases, survival culture and human behaviour, technological development, and metaphysical factor.

Environmental triggers, such as climatic change like glaciations, are the most explored within the common evolution theory. Threat from other organisms, such as diseases, has also been attributed to past species development. It is now emerging that besides the two triggers, survival culture, which includes economic ideologies and governance, has played a major role to propagate technological development. On its own, technology enforces the theory of evolution that is based on *coexistence for survival.* The weakest of the species are saved to support propagation to the next level thanks to technological advancement in medicine and production. Finally, collective human behaviour in itself, especially *euphoria,* is a factor that has led to wars and even sometimes threat to extinction.

Environmental Factors

Environmental factors include both physico-geographical and climatic conditions. The physical environment and climatic conditions determine the scope of resources availability, the nature of the ecosystem in terms of sustainability of the species, the extent to which there is interruption of growth due to natural causes (e.g. volcanic eruption, drought, and whether patterns). It is not therefore by accident that the better part of the evolution process of hominins took place at a given location. This location (i.e. Eastern and Central Africa) was environmentally the most ideal for the survival and eventual take-off of the *Homo sapiens.*

Climatic and geographic transiential triggers

The atmospheric and surface activities of the earth that occur naturally have in some instances resulted to disaster and calamities leading to threats to the life of man and property. For over a million years of existence, the hominins have lived and evolved in the regions of East and Central Africa devoid of earthly disturbances, such as earthquakes, volcanoes, tornados, wildfires, and floods among others.

Until the great migration of the groups of *sapiens* (some of whose migratory route ended at northern Africa, while others crossed to the Mediterranean, Asia, and Europe), the hominins had lived in a habitable environment with less environmental calamities or none. Unfortunately, the harsh climatic conditions and the highly intensive earthly disturbances affected the *sapiens* population that migrated out of Africa. Nonetheless, these migrations, not caused by geographical or climatic factors, from Africa's most habitable zone was caused by power struggles between various groups, where the most superior and advantaged displaced the disadvantaged ones and remained at the most favourable locations, driving weaker or inferior groups in search of other places as a normal animalistic behaviour for survival.

These types of displacement from previous habitats have happened across history, and to mention but a few, there were the displacement of the original American population and aborigines by the whites during immigration to America, the Arabs and *Carthaginian* displacement of the local populations in North Africa, and the displacement of the Masais in East Africa by the British directly and through treaties among others. The current habitats of human communities are a factor of competitiveness of communities over others as far as the most habitable surface of the earth is concerned. But that's beside the point.

At the final settlement areas of the migrated groups of *sapiens*—that is, the Americas, Europe, and Asia—further displacements arising from natural calamities coupled with harsh climatic conditions occurred or resulted to some adaptation. First, climatic conditions, such as prolonged winters or high temperatures in Europe and Asia, required more than mere common sense to survive. The first cycle of exposure to such climatic conditions in Europe and Asia must have devastatingly affected the initial group, leading to high mortality of the first generations and to adaptation by the subsequent generations. These prolonged winters and inconsistency in relationship between day and light, night and darkness, and vice versa led to, among other adjustments, innovations in the measurement of time and more ability to predict the future, the production of surplus food and clothing for cold weathers, and the development of structures resilient to harsh weather. The initial high mortality rates further leading to higher population growth.

My initial experience (contact) in Europe was in 2007, when I was invited to the Tallberg Forum and the new-leader training conducted in summer by the Tallberg Foundation in Sweden. It was also an opportunity to visit my family friend Miss Ira Sandberge in Stockholm, so I informed her about my travel itinerary, and she agreed to pick me up at Stockholm Airport. So in the chilly morning of 5 July 2007, I boarded the Kenya Airways flight at 8 a.m. at Jomo Kenyatta International Airport, Nairobi, connecting through Amsterdam, Netherlands. The trip to our first destination took eight hours. Prior to this trip, I had already been informed that Amsterdam was approximately three hours ahead of local time; therefore, I was expecting that I would arrive in Stockholm at night (i.e. 7 p.m.). Indeed, on confirmation by the clocks, it was 7 a.m. in Amsterdam when we landed. I then decided to make a phone call to Nairobi just to confirm that it was actually evening since the position of the sun in my new location was contradictory to my normal time in Kenya. Had I been in Kenya, the sun would not betray me; it would actually be midday or roughly twelve noon by the clock. As I got on board the next connecting flight, which was KLM from Amsterdam to Stockholm, I thought things would change and I would experience the night's darkness as I had expected. Three hours later, on arrival at the Stockholm international airport, I was much more astonished to see daylight. It should have been around 10 p.m. at night. When I finally met my host, she informed me about winters and summers. The winters are so cold that some people tend to be inactive, and summers are so hot; also, darkness can run for days and the daylight for even up to two weeks consistently. I definitely concluded that one has to be innovative to survive in such locations.

Secondly, most of Europe and Asia are prone to such geographical risks as earthquakes, volcanoes, landslides, floods, tornados, tsunamis, etc. These have been overcome over the years through avoidance or by creating physical and social mechanisms for survival, such as mystification, development of warning signs, and identification of safe residential areas.

Survival Behaviour and Culture

The survival culture is reflected in all spheres of activities undertaken by man and determines the behaviour of human beings as individuals or as communities. This relates toow they fend for food, how they share resources, and how individials in communities behave internally and with others. Humans' survival culture is *communal* as an inherent strategy to enhance their survival capacity as communalism gives them the synergetic advantages to reduce threats through joint action in addition to creating a support mechanism to ensure survival through the most disadvantageous circumstances. Hominin organization and survival capacity is therefore proven by the ability to socially assign roles, organize family, create threat detection surveillance structures, and set out rules guiding each communal group that has distinguished itself from others.

With developed rational and physical mechanisms, humans' survival culture has been regulated by inherent fear and uncertainty, myths, and at a higher level, laws. Whichever it is, the culture and behavioural overview of each group is a reflection of its social ideology, which is a survival mechanism. It is therefore possible to establish unique characteristics for each human group depending on how it is organised. Such organisation could either be progressive ; fast or slow, regressive, leading to variations in pace of social, economic, technological and political growth as it has been witnessed all over the world.

A summary of all political and economic ideologies advanced by scholars of the nineteenth and twentieth centuries all boil down to the description of the hunting culture of man. It is the form of hunting which includes sharing of the kill that has changed in different times and locations. Traditionally, in the archaic stage of the development of *sapiens*, a set of hunters would fend for the others, especially the females and children, and each hunter designated to provide for his family would take a portion of the kill. However, there are cases where the most advantaged either physically or in swiftness would take advantage of their abilities or have the largest share of the game no matter their need. This is the beginning of rules.

This implies that capitalism, socialism, or any other political ideology practised today within different political environments are just mere expressions of collectively agreed hunting rules within different states or by different regimes. The forms of governments are only a means to enforce, regulate and support hunting habits. It is for this reason that governance forms a critical indicator of evolution as it becomes the enforcement tool for adaptation in the next level. As agriculture and its associated technology developed and spread across the globe, man started to become largely agrarian resulting to a predictable and most reliable method of survival resulting to a proximate uniformity in human social behaviour.

Technological Advancement

The use of tools and the progressive advancement of these tools critically contributed to the evolution of man. As the erected man became slower than the animals they depended on (e.g. antelopes) for food, the sophistication of the tools improved, thereby enhancing his ability to survive. Apart from hunting purposes, man has used these tools to protect himself from other predators and to defend himself when there is conflict between different human groups.

The handy man, or otherwise *Homo habilis*, used simple tools, such as stone, for cracking nuts and hunting. Then as the evolution process progressed, better tools were developed, and eventually the archaic *Homo sapiens* devised a spear and arrow, increasing efficiency for hunting and thus expanding human diet with bigger kills. In terms of human-to-human interactions, it has been confirmed by both historians and archaeologists that man was largely at war with himself. An unfamiliar group could fight and conquer another's territory, especially to expand hunting grounds. The development of technology for defence (i.e. war) enhanced man's hunting capabilities from physical fights to the use of spears, then bows and arrows, and eventually the invention of guns and the use of gun powder. These weapons invented for human conflict ended up enhancing man's dominance in the environment.

In response to the environment, technological capabilities determined the nature of habitats and survival in different environments. Further cultural interactions are facilitated by the development of simple transportation modes, such as dhows, which facilitated the early Mediterranean trade between continents and the spread of religion and cultures. In summary, technology as a trigger for evolution has been progressing over the years, and its effects is most visible in the next chapters as we discuss the new species.

The early historical settlement areas like Egypt and Mesopotamia (3000–8000 BC) cannot escape the mention of initial technological developments of the *Homo sapiens* that facilitated human advancement to date. In Mesopotamia, which is the present-day Iraq, the flow of the rivers Tigris and Euphrates enhanced fertility that could sustain early agricultural practice in the Arabian Gulf. Additionally, flooding occasioned by heavy rains in the Zagros Mountains facilitated water for irrigation.

In Egypt, an initially hunter-gatherer community that found the shores of Nile habitable transformed into agricultural communities, taking advantage of the fertility of the soil and the constant supply of water flowing from the Nile. As the Nile community became agriculturally dependent, more tools—mostly farm implements, such as hoes, sickles, sticks, and later, ox-drawn ploughs—were used to increase production. During droughts or seasons when the water levels of the river went down, shadoof, a wooden device for lifting water from the river into canals, was invented and used. It is a large pole that swings and is supported by posts with a heavy load on one side and a bucket on the riverside. The weight of the load assists in pulling the bucket upwards for it to be emptied. Similarly, in Mesopotamia, surplus agricultural production led to the building of storage facilities, such as grain stores, and wheeled carts for transportation of farm produce.

For fishing communities inhabiting riverbanks, lakeshores, islands, ocean and sea coasts, the rafts, canoes, and boats were developed to enhance fishing in larger areas.

Technological inventions that have changed and enhanced the evolution of *sapiens*

1. The wheel
It has been recorded that by 3000 BC, there is a possibility that the Sumerians in Mesopotamia had invented the wheel. Use of rollers and then carts to carry surplus agricultural production increased the efficiency of transportation, making

movement of goods much easier. By around 2500 BC, wheels were used on horse-driven chariots for war. The wheel has progressively been useful to enhance human interaction as virtually all transport modes today utilize it.

2.   Pen and paper

The pen and paper are representative of tablets and are the early archaic tools used for drawing and conceptualization of alphabets that were the first forms of coding for storage of information. From drawings on caves, stone tablets, then paper, man was able to develop alphabets to encode and decode information that they intended to store and keep for the future generations. More information is now stored in audio and visual form.

3.   Printing press

In 1454 the German goldsmith Johannes Gutenberg was the first to construct a press that comprised moveable metal type, which, when laid over ink, could print repeatedly on to paper. The printing press increased the capacity of man to produce and store volumes of information and even share experiences widely. More bibles and other holy books were printed and distributed widely across the globe, and people were now able to share experiences and thoughts of others. Newspapers were printed for communication of events, making humans more aware of their environments and beyond.

4.   Gun powder

The gun powder became useful in increasing the efficiency of the armoury and duration of wars.

5.   Clothes and shoes

Early humans used hides and skins to cover their bodies from adverse weather conditions and for cultural purposes.

Humans who lived in climatic conditions that were not adverse did not therefore require clothing. Most parts of the African continent, Australia, and America embraced clothing as they began to interact with Europeans and Arabs from thirteenth century AD. But the significance of clothing is that by early 300 BC its need necessitated the great silk trade, which opened up the initial world trade interaction between Europe, Asia, and Northern Africa, leading to early cultural exchanges between humans. Today, clothing, including shoes, has become a mandatory cultural requirement globally.

6. Locomotive and motor vehicle engines

After the invention and spread of the use of wheels, donkeys and horses were used to pull heavy trucks of goods. After the design of the steam engine by the British engineer Richard Trevithick (1771–1833) and the eventual introduction of the railway locomotive, which could pull cargo-and-passenger trains in South Wales in 1804, further improvements were made to increase efficiency of rail transport, leading to the replacement of steam engines with diesel engines in 1892 by Rudolf Diesel. In 1883 Gottleib Daimler (1834–1900) invented the petrol-driven cycle, which was improved by Karl Benz to a tricycle with a petrol-driven engine. Eventually in 1886 and 1893, Daimler and Benz made four-wheeled vehicles. Today, both road and rail transports are the most efficient inland transportation modes.

Rail and road transport were most useful in opening up the world for mineral exploration, exchange of goods and services, spread of culture, and colonization of Africa.

7. Electricity

The invention of the electric dynamo by Michael Faraday (1791–1861) in 1831 was the beginning of the contemporary widespread use of electricity. Initially, electric power was generated through steam engines, then later developed to hydroelectric, geothermal, solar, and nuclear power. Through use of electricity, man is now capable of lighting, heating, and warming, using electricity as the source of power for communication equipments like mobile phones and computers, and driving engines for transportation and industries, making survival easier through work efficiency. Electricity still remains the driver of industrial development, communication technology, and innovations since its discovery.

8. Telephone

After Alexander Graham Bell's first experimental telegraph followed by the development of a transmitter and receiver in 1876, long-distance telephones were later introduced by connecting copper wires. By the 1950s, telephones had already been developed for vehicles. Subsequently, in 1973 Dr Martin Cooper and other co-inventors made the first portable cell phone handset and made the first cell phone call. Today, cellular technology has virtually taken over all telephone services. Communication has become instantaneous in addition to the capability of the communication gadget to be used in other functions, such as banking and information storage.

9. Internet

The Internet is a global information network system that facilitates communication by interlinking with compatible communication gadgets, such as computers. The Internet has facilitated a pool of knowledge and virtual experience for all humans across the globe, resulting in some cultural uniformity through cultural exchanges, interconnectivity, and increased non-indifference as exhibited today by the sharing of similar social medias across the globe. It has reduced the limitations that previously arose out of geographical confinements.

10. Radio

A Russian and the Italian Irish inventor Guglielmo Marconi saw the potential in the radio technology when they sent and received the first radio waves. Marconi sent the first transatlantic radio message (three dots for the letter S) in 1901. Since then radio became an important part of our daily life from listening to news bulletins to baseball matches, and even the invention of TV barely affected its significance.

11. Film and television

The introduction of the facsimile machine by Alexander Bain in 1843 and its later improvements in 1851 by Frederich Bakewel are seen to have been the first step towards the innovation of the television. Later on, with the Willoughby Smith's discovery of selenium as an element of photoconductivity in 1857, other improvements towards the development of the television were ushered in. After further improvements, instant transmission of images began, then came further technological improvements in transmission from stations to satellite, analogue to digital, and then the introduction of the colour television.

12. Artificial satellite

After the first satellite launch by the Soviet Union on 4 October 1957, several satellites have been launched into orbit. To date, more than thirty states have their space stations in orbit. Satellites support all communication, surveillance, transport navigation, and all virtual innovations that moved the world forward to interconnectivity and mutuality relationships both socially, economically, and politically.

13. Aeroplanes/aircraft

Aerial transport improved the intercontinental movement of humans by reducing travel time between both land and international waters. After this innovation, it now became easy for goods, services, and cultures to interact within a shorter time than it would have taken ships and any other previous mode of inland transport. Aircraft also took war to the next level, where the enemy is fought from the sky, as seen during the Second World War. The loss of life occasioned by these wars has now led to further development, leading to the use of drones.

Metaphysical Factor

Although it is difficult to derive logic from a process result triggered by unknown forces, nonetheless the possibility of *the unknown* causing change in human species cannot be ignored. The mention of a metaphysical factor here is for the recognition of the existence of metaphysical ideas in themselves. It therefore becomes almost impracticable to relate the intangible facts to physical change processes within this theoretical framework without reducing the subject matter to the arguments with *not substantiated* justification instead of the path of cause or triggers. Certainly, the other triggers are definitely within the human scope of verification and thus form the important fodder for which this theory flows. On the other hand, there are bodies or meteors in the universe which are beyond our scope but whose movement paths may affect certain aspects of the earth by collision or emission of substances that could have effects on the existence of plants and animals

Historical Milestones in the Development of *Homo sapiens*

A historical study of the milestones in the development of *Homo sapiens* reveals drastic advancement and carryover of cultural practices from one stage to another. A synopsis of European, American, African, and Asiatic history of the archaic stage of the species shows predominance in the development of settlements or communalism, establishment of mode of survival, and conceptualization of government. This is the prehistoric age to the *ancient era*, when there is evidence of metal use and adoption of agriculture and constructed habitats. Communal settlement became a consequence of communal organization and necessitated primitive forms of governance, mostly based on kinship.

Archaic to Maturity Stages of Evolution

The archaic stage covers the old, middle, and lower Stone Age as classified by archaeologists, including the initial stage of transformation from *erectus* to *sapiens*. In Europe, the *archaic stage* can randomly be tracked prior to and including the *Minoan*[5] civilization, *Viking Age*, as well as the beginning of the *Middle Ages*. In Asia, the same period accounts for the *Göbekli Tepe* in Turkey (dated to 10,000 BCE), the *Indus Valley civilization* (period 2600–1900 BCE), the Norte Chico civilization in the Americas, and the chiefdoms was Ta-Seti in Africa. Further, the histories of Mesopotamia (6000 BC–1100 BC), Egypt (3000 BC–700 BC), and Greece (circa 1000 BCE–323 BCE) up to the Late Antiquity Europe (fourth century to 900) represent the *archaic transiential stage* of the evolutionary development of the *H. sapiens*.

---

[5]   John Bennet, 'Minoan civilization', *Oxford Classical Dictionary* (3rd edn), p. 985.

Under these historical circumstances, the *triggers* for sociocultural development revolved around environmental conditions, food security, and threat from wild animals, diseases, and in some circumstances, other human beings. The proof of human response to these *triggers* informs the history of civilization before and during the ancient historical periods, with human communities experiencing more occurrences or higher intensity of the change triggers and exhibiting more complex ways of survival and detachment from nature. In other words, the more the *triggers*, the more human behaviour and culture adapt. This explains why the historical civilization models are different from continent to continent at the *archaic transiential stage*.

The next evolution stage is the *rational consciousness transiential stage* that is equated to the beginning of the *post-classical* era, dating before 500 BCE (in some cases), and through the modern up to the contemporary period. Some regions began to experience this stage before others due to variance in intensity of *transiential triggers* from one geographical habitat to another. The stage is outlined by conception and further development of religion to its modern form, trade, conquest, and finally realization and application of science. This period includes Europe (fifth century to fifteenth century), Scandinavia in Europe (793–1066), period of Five Dynasties and Ten Kingdoms in China (907–960), the First Punic War (264–241 BCE), and conspicuously, ancient Greece (collection of cities), where citizens were able to nurture debates that led to the birth of the most famous philosophers, like *Socrates*, *Plato*, and *Aristotle* among others. Traits learnt in the previous *transiential stage* were carried forward and *improved for coexistence* in this stage. Governance and mode of survival, including tools, were improved at the next stage as an upshot of the urge for competitiveness.

Increased experiences in climatic changes and threats to coexistence resulted to structural improvement of human habitat, which have now been revealed. Some were preserved as ancient ruins, such as the *Parthenon*, whose construction began in 447 BC when the Athenian Empire was at the height of its power; Dhāt Kahl in the South Arabia (eighth century BC to AD 275); and the obelisk at the Temple of Luxor, Egypt, among others. The interior of Africa is likely to have begun this stage in the fourteenth century at the beginning of intense slave trade.

The height of this stage was accelerated by the *Renaissance* that spanned the period roughly from the fourteenth to the seventeenth century, beginning in Italy in the late Middle Ages and later spreading to the rest of Europe. This period offered stimulus to scientific discoveries; increased human interaction, leading to wars; innovation of competitive modes of production characterized by intercontinental slave trade; discovery of locomotives; advancement of medicine; redefinition of religious ideology; domination; and colonization and liberation struggles in the African continent.

The explosion of activities in Europe during these historical periods thus awakened Africa and other continents by intensively exposing them to *transiential triggers*. These triggers include consciousness of existence of other people and races outside communal spaces, international trade, new technology, abolition of slavery, new religion, new governance structures, and colonization and Western education. The introduction of the Western religious belief system and education to Africa and other inaccessible parts of the world prepared the continent for global alignment to the challenges of the next period. One of the major occurrences that all continents were somehow prepared for is the introduction of a global culture as a driver for *cooperation* in the contemporary times going forward.

Finally, the beginning of maturity stage is reflected by the historical period of advocacy for great cooperation among states on humanitarian, political, economic, and cultural issues. This period begins after World War II, when United States president Franklin Roosevelt and Winston Churchill drafted the joint declaration by united nations at the White House on 29 December 1941, where twenty-six states signed a joint declaration of cooperation against nations which were termed to be savage, brutal, and seeking to subjugate the world (Hitlerism). This resulted finally to the formation of the United Nations organization on 24 October 1945, which is now an international cooperation.

Variations in Evolution

Differences in geographical experiences resulting from settlement in different areas and different adaptive cultural practices and advanced modes of survival were probably direct consequences of the differences in intensity of *transiential triggers* experienced by the various *Homo sapiens* groups according to their migratory routes and settlement areas during the first transiential period. A look at the migratory route of humans from their cradle around Eastern Africa reveals that the group that migrated up north could have experienced more variations of climatic conditions, more conflicts arising from narrow migratory routes, and more diseases resulting from climatic differences and geographical challenges in crossing rivers and deserts and finally settling in regions with unpredictable climatic conditions. The *Homo sapiens* group that remained in Africa settled within an environmental climate that was usual and had predictable hours of daylight and night darkness, temperate climatic conditions, predictable weather conditions, and a rich ecosystem requiring no (if not negligible) adjustments.

So as European *sapiens* were developing complex survival modes, habitats, and social units, the group in the African continent had no cause for similar adjustments. The rich ecosystem reduced the possibility of conflicts, the vastness enabled a weaker community to succumb or migrate further, and the predictable moderate climatic conditions allowed them to hunt and gather throughout the year, consequently reducing innovations for survival in Africa. With the variance of exposure explained above, a historical scan through the development of the two continents exposes drastic differences in mode of survival, governance, sociocultural behaviour, and technological advancement, making the African group just physiologically evolved yet with the general habitual similarities with the *Homo erectus*.

With these differences notwithstanding, there exist similarities in political organizations across the globe, though in different forms, pointing to communalism and human interaction as a uniform evolutionary trigger across the board. Thus as early as the archaic stage, there were empires or kingdoms reported both in Africa and Europe.

However, in terms of advancements caused by other evolutionary factors, not much development is expected from the African *sapiens*; all factors had remained constant. This explains why by the nineteenth century, some African communities (if not the majority) did not wear clothes, had no advanced medicine, had temporary shelter, and still survived as hunters-gatherers with minimum technological development and primitive weaponry.

In contrast, the species that had more experience with transiential triggers most likely realized *encephalization* to a different degree. Thus there is evidence that *Homo neanderthalensis*,[6] whose remains were found in Eurasia, from Western Europe to Central and North Asia had an

---

[6]   J. J. Hublin (2009), 'The Origin of Neanderthals', *Proceedings of the National Academy of Sciences*, 106.

average cranial capacity of 1,600 cubic centimetres.[7] The Neanderthal's cranial capacity is notably larger than the average (1,400 cubic centimetres) for modern humans, indicating that their brain size was larger. However, owing to their larger body size, Neanderthals were less encephalized;[8] nonetheless, this indicates the possibility of more intensive transiential triggers specific to brain size increase in human species towards that specific route and settlement area.

HOMO SAPIENS MIGRATORY ROUTE

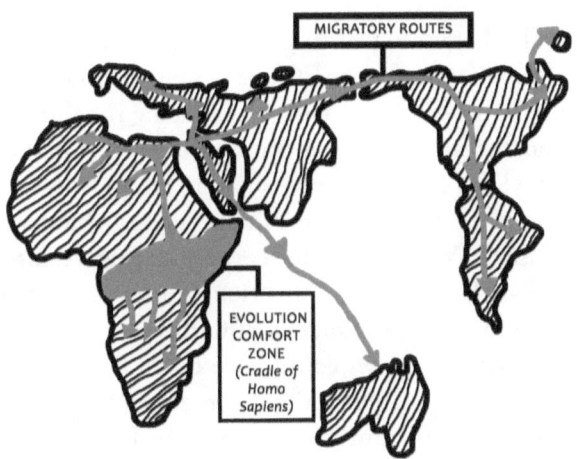

A map of early human migration.[9]

---

7   http://www.infoplease.com/encyclopedia/society/neanderthal-man.html.

8   *Homo neanderthalensis* (*H. neanderthalensis*) is a widely known but poorly understood hominid ancestor. <archaeologyinfo.com> retrieved on 24-5-2014.

9   Literature: Göran Burenhult: Die ersten Menschen, Weltbild Verlag, 2000. ISBN 3-8289-0741-5.

Variance of Lifespan Experiences and Transiential Stages

The length of exposure to lifespan experiences between Africa and Europe was different. Arguably, the duration it took for the *erectus* who lived in Africa to transform to the next evolution stage could have logically been the same period for which the archaic *sapiens* was to advance to *sapiens sapiens* if environmental factors had remained constant. The *sapiens* group that migrated to the north and settled in Europe, Asia, and other continents adapted to their changed environments, while the *sapiens* group that remained in Africa was retained within the same environment; their evolution triggers remained constant.

At the beginning of migration, exposure to different natural experiences for the European *sapiens* group increased their lifespan experience, thereby reducing the duration for evolution forward to the next transiential stage. On the other hand, the lifespan experience for African *sapiens* remained the same as they resided within the same habitat as their predecessors. While various carbon dating results have shown that the period of transition from *archaic sapiens* to *sapiens sapiens* is approximately 300,000 years, it is possible that the African *sapiens* could have taken even more years to exhibit *sapiens sapiens* traits or were even trapped within the *erectus–sapiens* transitional behaviour longer than their European counterparts.

The delay in the evolution process of the *African sapiens* from the archaic *sapiens* to *sapiens sapiens* was signified by the undeveloped social, cultural, and technological levels in Africa during its initial contact with Asia and Europe. The social activities that occasioned the transformation in Europe from the ancient to modern historical times escaped Africa. Apart from the European cold winters, this period is characterized by human conflicts, wars, and diseases, resulting to social and scientific innovations. In contrast, the vastness of the African continent, coupled with a wealthy ecosystem, reduced human conflicts. And as for diseases and environment, the African hominid had lived for millions of years (from before *Homo habilis* and *Homo erectus*) within the same environment, thus adapting to the adverse effects of tropical diseases within the East and Central African regions.

There was therefore no innovation nor was there a threat for their *non-adaptability*, resulting to stagnation of the evolution process and therefore delayed advancement from the archaic stage to the rational consciousness transiential stage. The historical and archaeological account for the ancient period in Africa is thus the account of geographical areas along the coastal boundaries with Europe and Asia. There is not much to account for in interior Africa save for scanty hunting tools that have been archaeologically discovered, indicating that Africa stagnated at the *archaic transiential stage*.

The following is therefore my take on Walter Rodney's justification: for the stagnation of Africa: When I first read Walter Rodney's (How Europe Underdeveloped Africa) piece giving justification on the underdeveloped status of Africa and attributing it to external forces, especially Europe, I was captivated by a false feeling that the situation in Africa (negative) is redeemable by charting our own path exclusively as Africa (some sort of self-determination) that could give a positive impetus to development. Rodney's argument blames Europe for its exploitation of Africa from the period of slavery and colonizationto neocolonization,resulting in the underdevelopment of Africa.

Now, all African states are independent and conscious of their sovereignty and therefore are self-determinant units. Institutions such as the Organization of African Unity are alive and functional, regional blocks are becoming stronger, and the African states have all become democratic.

Even with these functional units of renewed Africa, such as the states and OAU, it is now clearer to me that Rodney's piece was more of a counter-racial piece or response to racial discrimination encountered in Europe and expressed in Africa through the apartheid policies. With reduced racial feeling, the end of the era of self-discovery that characterized the emphasis on racial differences expressed through racial extremist behaviours, superiority complexes theorized by scholars, and inferiority complexes is over. We are now in a universal world upon which the propagation of human differences is stale. We are a world of non-indifference.

But for those still captivated by Rodney's ideas and other related ones, I reach out to you that development is an evolution trigger that eventually and necessarily equalizes every hominin group regardless of location on earth. As mentioned in the previous chapter, the reason for the stagnation of Africa in the archaic stage was more of strength and comfort than inherent weaknesses. I have already mentioned that the migration of *sapiens* out of Africa was based on displacement by the physically stronger group by then in a comfortable and more habitable environment. The next level of displacement was now ideological. For the migrated groups, rationality triggered by the harshness of environment had increased innovativeness, leading to much more artificial power. The power of the gun determined the strength of the group and their ability to capture and subject the other weaker groups.

What Rodney (who is representative of many) fails to mention is that a scan of written history of the world prior to the beginning of his assessment on Africa is scanty or limited recollection on African history due to lack of African recording ability. We know that ideas are improved by comparative conceptualization, but as we also now know,

Africa did not participate in the silk trade and never interacted with the other continents before colonization or initiated any contacts. Europe and Asia engaged or were colonized between themselves, engaged in wars that extinguished most of them, engaged in slavery for over 500 years (which is more comparable to Africa), fought more diseases, and struggled against colonization or imposition of other rules, yet they still developed comparatively better than Africa.

The idea of an African path to development interrupted by Europe has therefore no base. The reasons for development variations in relation to evolution have been explained. Hitherto the period of colonization and trade, which was more on slavery, the other parts of the world had received a brunt of the same (and even much worse) but got to a point of equilibrium. Contrary to Rodney's propergation of ideological self determination by Africans for African civilization and social-economic development the lone path of African ideology, ujamaa, espoused by Tanzania, has failed and is part of historical discussions for explaining Tanzania's economic stagnation during the period of the ujamaa ideology. The promoter of the ujamaa ideology, Mwalimu Julius Nyerere of Tanzania, admitted to its failure before he died, attributing this failure to the inherent behaviour of citizens and external influence. This is just but an example of many other attempts whose results end up with the realization that the path we take must always conform to a global euphoria or otherwise we fail.

Singapore is a good example of such a state that had accepted a global model for development and implemented a complementary framework with states that were comparatively more developed within the world's predominant economic and social ideology. In 1963, Singapore was on a par socially and economically based on economic indicators with the other states within sub-Saharan Africa. By 2013, it was incomparable to these African states that were still grappling with ideological confusion arising from identity crisis.

Therefore, if Africa was to successfully chart its own path to development, the course of evolution would drastically change. Speciation at this level of maturity may lead eventually to two distinct hominin species with one lower and another higher hominin.

Just imagine two children from the same father and mother—one living in the urban areas and another subjected to the forest life without any artificial support (i.e. clothing, technology). The one in the forest would start behaving like a monkey, ape, or gorilla in the long run, while the one in the urban area will develop the traits of humans as they advance in development. In the long run, the subsequent generations of the two in the different conditions will most likely have completely variant traits, or the lower one may become extinct due to the lack of coping mechanisms or may advance primitively in an animalistic manner that makes him more compatible to the natural environment than his counterpart, who would be adapted to the artificial environment that man has created today. The possible outcome is that he will become animalistic like other animals in the jungle and become part of a total package of the natural environment, whose survival is entirely dependent on the urban man. To survive, the man in the jungle will need interventions from the advanced man in the urban area. Further, he will be considered a lower-level hominoid by a successive generation of civilized humans if no intervention is undertaken to save him from that level. In which case, such interventions are at the mercy of his urbanized or civilized brother.

The *Australian Aborigines* are a good example of such an illustration; due to their hold on *natural* human behaviour resulting from their long lack of interaction with other cultures, they suffered devastatingly during colonization and were at the mercy of their colonizers to either survive or become extinct.

The question as to whether any *sapiens* group could have developed their own consciousness superior to another is neither here nor there as it is not possible. It is the interactions that enhanced or hybridized the capacity of each group's traits to conform to the same hominin standard and equilibrium of traits that we all have as humans today. No part of the world can therefore create a path different from the global path of humanity and succeed. Consequently, the evils or what has been largely seen as maltreatment of Africans before was part of a natural occurrence that eventually incorporated Africa to the future world of non-indifference of humanity. It is like introducing a bull to a herd of bulls. The new bull must be fought by all the bulls for protocol to be established, after which a harmonious coexistence begins within the herd.

Egypt

Whereas I have argued that the ancient history of the African coastal areas is an account of its interaction with Europe and Asia, Egypt is an exception. The explanation for Egypt being the first in 'civilization' is in the *Homo sapiens* migratory behaviour. The group that moved north up to Egypt faced the similar intense experiences with those that crossed the *Mediterranean* into Europe, including human-to-human conflicts, harsh geographical terrain, and unfamiliar diseases. The advantage this group had over all the other groups was that after all their experiences, they were the first ones to settle. Earlier settlement and threats from other migrating groups triggered the development of a social structure that ensured their survival. Geographical proximity to Asia is another factor that made the ancient Egypt unique to the rest of Africa in the pace of 'civilization' as it meant that fear of threats from other humans was multidirectional—that is, from the interior through the Nile, from Asia through the Red Sea, and from Europe through the *Mediterranean Sea.*

So those who settled in Egypt then had to improve their habitat by constructing stone walls to prevent entry of potential enemies, reducing potential conflict inland by farming crops as opposed to hunting, creating tools of war that would give them advantage over an invading group, and finally, creating an organized societal structure for governance and surveillance. Further, Roman conquest of Egypt in 30 BC adds to the earlier distinction from the rest of interior Africa.

Imposition of Rational Consciousness

Settlement, colonization, trade, and religion are the main activities that led to transformation of stagnant communities of the world from archaic to rational consciousness stages of human evolution. Indeed, it is on the basis of these that transformed social traits of humans are transmitted from one group of persons to another whose self-transformation is stagnant or slower. The Europe and Asian groups of *Homo sapiens*, having moved from archaic to rational consciousness stage of evolution ahead of Africa for reasons explained earlier, had settled in a geographical location with deficient natural resources and with minimum climatic comfort, and as the population increased, a search for more resources was occasioned by the scarce resources, and further migratory trends began with different groups migrating back to Africa. These activities propelled the development of the entire human evolution process as the group had higher levels of self-consciousness and a more systemized reality of their existence than the *sapiens* group that remained in the comfort zone.

Each of these activities are discussed below to demonstrate how, over years, the acquired traits and social adaptability mechanisms have been transfered by the *sapiens* that underwent intensive *transiential triggers*, making them socially equipped, universally conscious, and thus sophisticated. Where these activities occurred between different groups with superior sociocultural behaviours among themselves, the engagement was more *mutualistic*; both groups took a social aspect from another to develop itself as opposed to a relationship between inferior and superior sociocultural practices, where the latter mostly imposed, settled, and colonized the former. In the superior–inferior sociocultural engagement, the superior social behaviour became dominant, leading to the disappearance of the inferior.

Settlements

There have been several settlements arising from the migratory behaviour of human beings. Only a few historical instances have been chosen here to demonstrate the impact and extent to which some migratory trends affected the forward evolution of human, his sociocultural behaviour, and perhaps racial classification. We begin with the Phoenicians from Tyre, who by 814 BCE had moved back to the North African coast, settled, traded, and integrated with the native Berber population, leading to the establishment of the city of Carthage. Carthage later became a major trading city in the Mediterranean, largely through significantly trading with tropical Africa.

In Algeria at Djemila, northeast of Sétif, and Timgad, southeast of Sétif, settlements of veterans under the Roman emperor Claudius (41–54) were established. As a result of these settlements, the north exported grain, fruit, figs, grapes, and beans to Italy, Greece, and other parts of the empire.

In the first century CE, Christianity had spread to northwest Africa, but by the tenth century, the majority of the population of North Africa was Muslim. This was due to the migration and eventual settlement of the Arab nomads around the same period out of the Arab peninsula to North Africa, spreading the Arabic language and thus 'Arabizing' North Africa, which is Arabic today.

In the East African coast, some of the artefacts found in the early settlement areas of the Arabs, the Gede Ruins, indicate that even for the Arab sailors, there were tools and artefacts used that were borrowed from their contact with Europe, which as a result of settlement has had impact on the indigenous population to date.

Some of the artefacts, as indicated in the Gede museum, are listed below:

1. iron lamp (Indian)
2. iron scissors (Spanish)
3. Chinese coins with religious tittle Shao Ting (AD 1229)
4. fragments of Egyptian or Syrian cobalt glass.

Trade

Trade or commerce is the transfer or exchange of goods or services from one person to another for either other goods and services or money. It began in the form of barter trade, where goods and services were directly exchanged for other goods and services. According to *Peter Watson (2005)*, trade could have begun 150,000 years ago. Archaeological findings reveal that trade most likely began when different cultures began to interact. In the Mediterranean region, the earliest contact between cultures was of members of the species *Homo sapiens* principally using the Danube River at a time beginning 35,000–30,000 BC.[10] At least in all continents

---

[10]   D. Abulafia, O. Rackham, M. Suano, *The Mediterranean in History* (Getty Publications, 1 March 2011), ISBN 1606060570 retrieved 26-06-2012.

where human beings settled, there is evidence that trade began to take place among communities in different historical times.

In this evolution discourse, trade is an activity of coexistence that has essentially transmitted consciousness over time, beginning with the intercontinental trade. Most writers, including historians, have formed a general consensus that trade was an important way of transmitting culture and human social behaviour, sometimes creating new ones. The more societies traded, the more they developed socially and culturally. It is not therefore an accident that Egypt by the eighteenth century was completely different in socio-economic and technological development from the rest of Africa as evidence shows that it was part of the early Mediterranean trade by as early as fifth century BC. Egyptian interaction and earlier civilization meant mutual cultural exchange between it and Europe, with the socioculturally superior trading partner (Arabs) transmitting its culture. The remaining part of Africa and other parts of the world with similar characteristics of inaccessibility evidently traded internally within households with scanty possibilities of intercommunal trade discovered by archaeologists. Until the late fourteenth century when different tribes started establishing permanent settlement areas, there was little or no intercommunal trade in sub-Saharan Africa.

The major item for trade by various groups, especially between Europe, Asia, and Egypt was the silk trade, which resulted to the establishment of the Silk Route, which covered Europe, Egypt, Arabia, Persia, India, and China. Apart from trade in silk, other commodities and, more so, cultural exchanges occurred within the route, creating even more superior cultures in different areas. Trade on the Silk Road was a significant factor in the development of the civilizations of China, the Indian subcontinent, Persia, Europe, and Arabia, opening long-distance, political, and economic interactions between the civilizations.[11] According to Jerry H. Bentley in *Old World Encounters: Cross-Cultural Contacts and Exchanges in Pre-Modern Times*, the spread of Buddhism throughout Southeast, East, and Central Asia in the first century AD is attributed to the Silk Route.

Essentially, the regions that were left out of these trade interactions stagnated and did not therefore advance socially, economically and technologically as compared to the rest. Inland Africa for instance, began to interact mainly through the coastal region increasingly with Arabs after the fall of the Roman Empire around fifth century AD. However, this interaction was limited to the Arabs' objective of controlling the trading ports of the seas and oceans, thus hardly having much cultural interactions with inland Africa.

---

[11] Jerry Bentley, *Old World Encounters: Cross-Cultural Contacts and Exchanges in Pre-Modern Times* (New York: Oxford University Press, 1993), 32.

SILK ROUTE
A Sketch Illustration on how the Silk Trade
around 210 BC

Main Silk Route

Olker Route / Trading Routes

A map of the Silk Road trade route.[12]

As the most well-known overland trading route of ancient civilization, the Silk Road grew under the Chinese Han dynasty (202 BC–AD 220) during the first and second centuries AD and connected the Yellow River valley of China to the Mediterranean Sea.

The relevance of the social development among the *sapiens* interaction through the Silk Route is the complexity of social

---

[12] http://silkroutes.net/Orient/MapsSilkRoutesTrade.htm retrieved 17-11-2014.

Vadime Elisseeff, *The Silk Roads: Highways of Culture and Commerce* (UNESCO Publishing/Berghahn Books, 2001). ISBN 978-92-3-103652-1.

characteristics that was facilitated by these interactions and how it gradually spread across to transform all human races socially.

Until the beginning of the Age of Discovery in the fifteenth century and the explorations sponsored by Henry the Navigator, knowledge of the African interior was limited to the other world as the Africans in the interior too had scarce idea of the existence of other continents and humans. By then, slave trade began flourishing as religious wars between Muslims and Christians yielded slavery. In 1452, upon the approval of Pope Nicholas V, Portugal legitimized slave trade.

Much more has been written about slavery; however, the significance of this exploration and decree is that the two events found the *sapiens* living in interior Africa less developed socioculturally and technologically. They had not been faced with any conflicts arising from interaction with other races and neither had they established a strong rationality for self-discovery. In other words, they were still archaic and habituating within the *Homo erectus* cultural environment. Evidently, they had to be the target of slave trade.

But the critical evolutionary aspect of the slave trade, especially in Africa, was that it was the beginning of interior sociocultural relationships with more complex cultures and awareness of existence beyond the African communal environment for the African *sapiens*. For the enslaved minority who survived or were not castrated, it created further consciousness of the unity of the species. The masters of the slaves were able to recognize similarities in human characteristics and humanity, and in some instances, cross-racial reproduction began, leading to the campaign for abolition of slave trade.

Religion and Colonization

The history of the world's most well-known religions—that is, Christianity, Islam, Hinduism, Buddhism, and Judaism—and how they have affected the evolution of mankind correlated to their spread. The spread of religion introduced sociocultural practices to the people converted, causing them to abandon indigenous religions that had been adapted to local conditions. However, the most multicultural and cross-continentally spread religions that have accelerated rational consciousness over the years are Christianity and Islam. As mentioned earlier, the spread of Islam by the Arabs had by the tenth century led to the Arabization of northern Africa. In the coastal Swahili areas, the locals who were integrated with Arabs became Muslims and were therefore able to learn Arabic and trade effectively with them.

The Berlin Conference of 1884, which regulated European colonization and trade in Africa, is usually the starting point of the conquest of Africa.[13] European exploration of the African interior began in earnest at the end of the eighteenth century as a result of technological advancement in Europe. European industrialization improved the capacity to explore inland Africa with the development of transport and communication technologies. The steam engine led to the development of railway lines that opened up transportation in Africa and began Africans' exposure to the social and technological development of the world.

Eventually, the invasion, occupation, colonization, and annexation of the African territory by Europeans further paved the way for Christian missions due to the

---

[13]   Patrick Brantlinger, 'Victorians and Africans: The Genealogy of the Myth of the Dark Continent', *Critical Inquiry*, 12/1 (1985): 166–203.

secure environment that the establishment of the colonial government created for them. On the other hand, the presence of Christian missionaries supported the colonial regimes through education and recruitment of human resources that were to support in management of the colonial government. As a matter of fact, the African nationalist movements that led to independence were led by Africans who had got education and even worked in the colonial regime.

Colonization and religion thus provided the stimulus for the *Homo sapiens* in inland Africa and in any other part of the world that had stagnated to move a step forward from the archaic stage to the rational consciousness stage. It is probable that the experiences of interracial interaction between Europe and Asia during or around the first century (during the silk trade) that led to rational consciousness started being felt more intensively by inland Africa 1,800 years later. By the end of the decolonization of Africa and any other colonized region of the world, the colonizing group proliferated their culture into the colonized population, imposing *rational consciousness* and consequently sowing seeds for a thriving future global community.

CHAPTER 3

COEXISTENCE FOR SURVIVAL

As a departure from the *natural selection* theory that focuses on evolution on the basis of the most advantageous to survive and propagate the next level of its species, this theory is an attempt to prove that evolution has been majorly based on *coexistence for survival* and that coexistence is the driving explanation for the development of the human species to the next level. Further, justification for the extinction of other species can then be explained on the basis of their inability to coexist, which is explored in the next chapter, as opposed to their inherent physiological ability to cope with or adapt to the changing environment.

Humans have socially adapted through the evolutionary stages, exhibiting three coexistence modes—namely, *predation, parasitism*, and *mutualism*. The modes are social and biological relationships have determined human to human and human behaviour socially and humans and other animals biologically. So far, being *omnivorous*, human beings have acquired the ability to utilize all modes of survival, placing them within the most advantaged position in a biological environment or ecosystem as compared to other animals.

Each habit of coexistence is enforced either singularly or jointly with another, depending on the *transiential stage's* evolutionary adjustment requirements arising from the predominant transiential trigger.

Predation

In this context, predation is a survival relationship between animals where one has to be killed for the other to survive. The killing may be for either direct consumption or socially motivated to increase space for another survival requirement. In ecological science, *predation* is defined as 'a biological interaction where a hunting organism feeds on another that it attacks'. Predation is predominant within the *archaic stage*, when humans are still not able to relate their relationship with others animals. This stage, as explained within the previous chapter, is characterized by communalism and human predatory behaviours, such as hunting and even savagery that exterminate others of their own for hunting space and, in some cases, for food consumption.

Historically, what encouraged the predatory behaviour of humans was their assumed belief that nature offered plenty and was self-replenishing. The existence of plants or animals had not been accounted for within a clear rational thought to enable human understanding that these food resources could perhaps be limited. In difficult climatic conditions that occasioned starvation, humans fed on human flesh, which was rare human to human savegery, offering little chances of survival. Additionally, the development of religion also contributed to the predatory behaviour between man and other animals but limited it to some extent between humans themselves with the exception of instances of human sacrifices. Until religion blended with scientific facts of nature, man's belief in provision by metaphysical forces of nature could have led to near extinction of some animal species and intolerance in human relationships, resulting to extermination. As explained before, man carries with them some traits despite stages of development within the evolution stages. In different form, the contemporary predatory behaviour is sociocultural in form and is exhibited between humans who exterminate others to satisfy their sociopolitical, cultural, or religious reasons. In religion today, it is exhibited by some religious fundamentalism.

Predation is, however, not the best form of coexistence as it leads to extinction of some species,creating a gap in the ecosystem that eventually affect all species. The African elephant (*Loxodonta africana*) and the Sumatran tiger (*Panthera tigris sumatrae*) are threatened to extinction. According to the World Wildlife Fund (WWF) organization, the Sumatran tigers are critically endangered, while the African elephant is vulnerable. Poaching for trade is responsible for over 78 per cent of estimated Sumatran tiger deaths, which consists of at least forty animals per year.[14] The African elephant, like the Sumatran tiger, has experienced commercial poaching (ivory) and reduced habitat due to the expansion of human habitat, resulting to human–animal conflict.

In the case of humans, extinction can only result from social predatory behaviour arising from religious fundamentalism and conflicts arising from resources and political competition among others. Thus, global institutions have been put in place as a social mechanism for preservation of both humans and any other endangered species that would be extinct by now. The contradiction in the efforts to eliminate predatory behaviour is a similar action to that of the predatory man, which is also self-extinguishing.

---

[14]   https://www.worldwildlife.org/species/directory?direction=des c&sort=extinction_status retrieved 11-11-2004.

Thanks to increased rationality, majority of humans understand the utility value of other humans and animal species and therefore must necessarily exist with them, thereby increasing the chances of continued existence of the next generation. Thus, there are efforts to ensure the continued existence of the black rhino (*Diceros bicornis*), cross river gorilla (*Gorilla gorilla diehli*), African elephant (*Loxodonta africana*), and Sumatran tiger (*Panthera tigris sumatrae*), which have been declared endangered species or threatened by extinction.

Parasitism

With further realization that more benefits are derived from a synergetic engagement and further awareness on joint challenges of survival and appreciation of diversity, human coexistence modes tend towards mutuality.

Let's begin by defining *parasitism* as 'a relationship between species where one species benefits at the expense of the other'. This definition includes typically smaller organisms, like viruses and bacteria. The intention of the parasite is to survive as long as the host lives. In this relationship, the parasite does not necessarily care about the welfare of the host and therefore, in most cases, affects the host adversely. For example, the *Plasmodium* bacteria, which causes malaria, enters the red blood cells of a living human to get food for its own survival. Such a relationship is harmful to the host since the host loses some of its vital components; this makes them unwell, sometimes leading to the death of the host. Historically, Parasitism was well exhibited by humans during the slave trade in the beginning of the fifteenth century; Africans did not participate, and in many instances, the African people were simply the victims (Walter Rodney 1973). The African slave was the host, while the master was the parasite.

The modern up to contemporary manifestation of this survival habit is exhibited in unfair trade practices, poor labour relations, or exploitative remunerations.

Mutualism or Symbiosis

By definition, mutualism is a relationship between organisms in which each organism benefits. In *obligate mutualism*, the species are interdependent and cannot survive without each other, while in *facultative mutualism*, the interacting species derive benefits without being fully dependent.[15]

Mutualism is exercised at different levels—namely, *individual*, *societal*, and *global levels*. The individual level is the reproductive level, and the societal or communal interaction starts from kinships, then tribes, and up to the state. Finally, the global level is multiracial, multicultural, and multinational and is the most advanced level of human coexistence that creates a conducive succession ecosystem for the entire human race to the next evolved species. The three levels intertwine for the benefit of the contemporary hominid evolution process; thus, be it global, societal, or individual interdependence, the main result is to propel the infinite evolution process of the human species.

While on a social tour to the Gede ruins in Kilifi County formally in coast province, I noticed how coexistence between Arabic culture and the Mijikenda communities led to evolution of a new community and language, Swahili. In early fourteenth century AD, it is approximated that seven hundred Arab men set sail and docked at the Kenyan coastal shore in Malindi. The natives were so welcoming, and so they decided to settle around the Gede area, where they intermarried with the locals. The significant evolutionary

---

[15]   http://www.thefreedictionary.com/mutualism retrieved 30-10-2014.

details of the Gede lesson are the results for individual, multicultural, and multiracial coexistence that led to a new community and culture that is in existence to date. That is mutualism.

A further look at the artefacts found within the ruins reveals some earlier interaction with the Chinese culture. Spoons and plates used by these Arabs living in Gede were made in China. That was facultative mutualism. Basically, any trade arrangements undertaken by communities from medieval to modern and contemporary times—be they multicultural, multiracial, interstate, or individual to individual—where there has not been exploitation or degrading treatment between humans is the best form of mutualism.

Mutualism as a Necessity for Human Survival

The more the society develops, the more individuals become dependent on one another converse to expectations that development should generally lead to independence. The more independent the society of man, the most likely the society develops predatory or parasitic behaviour that is a threat to the human evolution process as history of man has shown. Indeed, a single social unit so organized characteristically behaves as an individual in its irrational form and tends to have traits that are retrogressive to it and other societies. These retrogressive traits are self-extinguishing to the human species both within the organized (developed) units and outside their jurisdiction. It is for this reason that the two world wars fought so fiercely by mankind beginning 1914, world war 1 and beginning 1939 world war 2 were fought by the most developed nations of mankind among themselves.

Thus the self-destructive traits of each organized and developed state or community could have led to extinction of its own kind in some parts if not the whole world.

So assuming all organized states or communities were independent and not interdependent through any community of nations to stop the Nazi Germany regime, the Rwandan genocide, the Israeli–Palestinian extinction threats, and most likely the global threats of development of nuclear weapons, terrorism, and dictatorial regimes of the independent nation-states among others, societal mutualism would not be of benefit to the evolution process, yet the contradiction is how the perpetrators are socially attracted to these primitive habits. As affirmed earlier, social predatory behaviour is in other words referred as savage behaviour. And collective in thought (or mobocratic), man is easily swayed to his basic instinct of predatory behaviour, which is self-destructive. Humans have therefore developed global obligate mutualism to protect the society and individuals from threats.

In contrast to Darwin's assertion that 'natural selection acts solely through the preservation of variations in some way advantageous, which consequently endure', events over the last 100 years have shown that unlike the later generalization that drove the natural selection discourse, humans have adaptive tendencies that make up for certain weaknesses that would otherwise make them disadvantaged in the selection process within the state of nature both individually and collectively.

For example, at an individual level, a case of blindness results in the enhancement of other senses, or when one arm is physically inadequate, another arm develops to compensate for the two-arm capability. At a collective (community or global) level, the possibility is higher that the weakest of communities will marshal their unity to fight the strongest and most advantaged singular community that threatens their ideals and existence.

Again the most advantaged is always the most comfortable, with lower coexistence instincts, and is therefore ordinarily the most likely endangered and in constant threats from conspiracy of the weak. Because each single unit would individually seek protection from the strongest when threatened by a relatively stronger unit, there is predisposition to cooperate with the strongest and the most advantageous. Therefore, to reduce the fear of uncertainty, increased cooperation among the weak themselves and between the strong and the weak has been necessitated.

Sexual mutuality

At the individual level, humans interact for procreation (between males and females) and companionship. Mating is the most fundamental cultural and biological relationship that has propelled the evolution of the human species and supported its urge for survival, resulting to reproduction, nurturing of the young, and the first-line protection of the hominid during its lifetime. So even at this very basic level, the society and community of societies (global level) dictates the individual behaviour by creating rules that culturally guide individual reproductive behaviour. Thus, the society has organized this mutuality of two mature people of opposite sex as marriage and has consequently assigned each of these units the responsibilities of nurturing and protecting the young ones even though the responsibilities culturally assigned coincide with natural instincts. Nevertheless, mating (even out of a non-culturally classified mutual relationship) under favourable biological conditions by humans of opposite sexual orientations results to reproduction.

Among the Masai community in East Africa, any man within the same age set was allowed to have a reproductive relationship with anyone's wife in their age set. This culture allowed the Masai community to increase their population amid shocks of a myriad of diseases, droughts, and environmental challenges that would have reduced their population and is an expression of societal-level mutualism.

The other driver of coexistence is fear, which compels single individuals into companionships such as marriage, family ties,kiships e.t.c, as a source of social safety nets. This occurs even inside socially secured communities since the psychological environment described by Thomas Hobbes in *Leviathan* (1651) as a pure state of nature where man is at war[16] and where 'life is solitary, poor, nasty, brutish, and short' (*Leviathan*, chapters 13–14) remains ubiquitous. Man does what would make him preserve his life, which includes his own future gene lineage. This intrinsic natural characteristic is one of the compelling forces on which family bond is founded even in contemporary times. Inherently, at an individual level, even the most civilized individuals live in constant fear, mostly of unpredictable occurrences that makes existence uncertain. This fear in *Hobbesian* description of life is what drives the transformation of man favourably towards the most complementary mode of survival for his own and successive species.

Apart from social and psychological evolutionary adjustments, the major biological advantage is the pleasurable mating or sex, which has made it possible for men to establish relationships with the opposite sex for recreational purposes, thus making it possible for conceptions that are secondary to the objective of sexual intercourse. Indeed, there is possibly one child or more conceived in a family as an unpremeditated by-product of recreational sexual activity. Certainly, the basis for modern innovations on birth control, coupled with the statistics of deaths related to sexually transmitted HIV/

---

[16] Thomas Hobbes, *Leviathan*, ed. Edwin Curley (Hackett Publishing, 1994).

AIDS, is the consequence of this biological adjustment that drives human urge for individual-to-individual relationships.

At the societal level, individuals establish a higher bond that defines a collection of different individual-level coexistence units. As family units are formed as a result of individual relationships, so are societies formed by different families establishing a unifying bond, thus making individuals and families the building blocks for mutualism.

CHAPTER 4

HUMAN SUCCESSION ECOSYSTEM

A succession ecosystem is the natural and social environment within which species (humans) interact to stimulate propagation and survival of their successive generations. Characteristically, the succession ecosystem can be discerned by discussing social and natural adaptive behaviour that man has developed in order to deal with the challenges of his survival or transiential triggers.

As I mentioned earlier, in terms of external aesthetical changes that are physiological, not much is expected to happen at this stage, meaning tail or additional legs are not expected to be seen. However, there is likelihood of a lot of biological and structural changes in variance between the *Homo sapiens sapiens* and the contemporary man which has been directly caused by the artificial modification of the environment to enhance man's adaptability for the next level of evolution.

The answer to whether man has actually evolved is therefore systematically dealt with in the next chapter by demonstrating the results of these gradual environmental and social changes that have preconditioned the hominids' adaptation over the years. This chapter deals with the description of the social and environmental measures that man has radically adopted over the years to create an ecosystem that facilitates his succession to the next level of hominin.

Certainly, this ecosystem is comprised of natural environment and social environment. The natural environment includes the climate and the physical and geographical structures and habitat within which man lives. The social environment includes survival modes, governance, education, technological advancements, and cultural behaviours that include recreational activities.

The Natural Environment and Climatic Conditions

The aspects of the natural environment that have affected the human evolution process include geographical forces, atmospheric conditions, and flora and fauna. For example, in some traditional African societies, forests that had medicinal plants were mystified as a way of ensuring their preservation. Only medicine men were allowed to enter them. This helped to maintain the constant supply of medicine for the communities by reducing human activities that would endanger the forests. This is further emphasis that social-behaviour change is a human's adaptive behaviour for survival.

Naturally, like other animals, man coexisted with nature in forests as forests provided him with a rich environment for hunting and gathering and shielded him from harsh climatic conditions, rains, etc. As man evolved, man modified his habitat, making him able to survive out of forests, or probably reduced the dangers experienced in interaction with other predatory animals in the forests. Thus the forest cover began to reduce since the basic resource for development of habitat was trees. Technological advancement and mandatory use of fire fuelled by wood virtually drove human habitats from forests.

Exceptionally, some communities—such as the Ogieks in the Rift Valley, Kenya, and the Pygmies of the Ituri Forest in northern Congo, who have continued to be hunter-gatherers—live in forests to date. However, the increasing human population and its needs, coupled with technological development, have within the short period of existence of the *Homo sapiens sapiens* affected the climate and the natural environment's characteristics. While some scientists have documented how these changes have affected man today, the realization that the effects of man's activities on the environment, whether negative or positive to man himself, are an end to itself as an evolutionary advancement has not come forth. These natural environmental aspects revolve around geographical and climatic conditions, forestry, human habitats, animal husbandry, agriculture, and climatic conditions.

Mitigation Measures
for Geographic and Climatic Hazards

Long before human population expanded, the intensity of the negative outcomes of adverse geographical and climatic conditions that existed were predictable and, to some extent, avoidable by humans. Adverse geographical and climatic conditions have been confirmed to either occur naturally or as a result of human activities. Over the last 200 years, human populations have expanded, leading to indiscriminate settlement and development of human habitats in areas that were previously not liveable.

Occasioned by the Industrial Revolution in Europe, mankind intensified the exploitation of natural resources through activities such as agriculture, mining, delineation of forest lands for settlements, industrial activities, wars, and deforestation for timber industries among others; these led to adverse weather conditions, increased desertification, flooding incidences, pollution of air, depletion of the ozone layer, emergence of diseases related to climatic changes, etc. Further, as mentioned in the previous chapter, there are geographical behaviours which are characteristic of the earth and have potentially calamitous and disastrous outcomes that man has had to live with or deal with over the years. These majorly affected the immigrant *Homo sapiens* (i.e. European and Asian *sapiens*) within their locations of settlement as experiencing more plate tectonic movements leading to earthquakes, volcanic eruptions causing pyroclastic emissions and landslides, and adverse climatic conditions arising from unpredictable weather patterns just to mention but a few.

With all these life-threatening conditions, humans innovated survival mechanisms to continue habitation in areas with high potential for naturally occurring earthly disturbances and corrective actions for those intense geographical activities arising from man's activities through the united efforts of man exemplified by the environmental charters of the United Nations, such as the Kyoto Protocol. I will not discuss these environmental efforts, but it is necessary to mention that these are measures for the preparation of a sustainable environment for the future generations, which serves the purpose for the creation of a conducive succession ecosystem for *Homo x*.

I would have thought that the other alternative for humans would be reverse migration to the original habitat, which is in the East African region, as a risk avoidance mechanism. However, re-settling over 7 billion of the current population beats the logic of evolution, especially after further figuring out how devastating a disease like *ebola* would be to mankind and concluding that such a tragic scenario has a magnitude of devastation that would most likely render man to extinction. So I abandon the idea that we should all congest ourselves in the safest part of the world with a clear understanding of the advantages of the dispersed settlement of mankind.

I will use two disasters to explain the extent to which man has gone to improve or lower the risks arising from natural disasters—one natural and one man made.

a) Earthquake-prone areas

Data on occurrence intervals of earthquakeshave been established to assist in predicting the next occurrences; measurement tools for seismic intensity were improved for the detection of occurrence and magnitude. The seismograph is designed to record levels of disturbance by registering the maximum deflection of seismic waves from their amplitude recorded in the *Richter scale*.

Another measure is earthquake engineering, which is directed towards the design and construction of human structures that would be less affected by the amplitude of seismic movements or earthquakes so that buildings designed and constructed in consideration for earthquakes are solid and stiff enough to resist crumbling. Additionally, as a measure to protect lives, governments in earthquake-prone areas have enforced building codes and established seismic zones for continuous geological studies through risk information and insurance, disaster preparedness, and regulation of human settlements.

b) Human-caused disasters

The major contributor to environmental changes today is human beings. To develop within the evolution process, man performs certain activities which affect the environment yet are necessary for development.

1. Agriculture is the major cause of deforestation in the world. It has negative impacts on the environment by causing soil erosion, depletion of some plant species as a result of clearing of natural forests, and reducing water catchment areas, which are useful for the regulation of the weather, resulting to desertification.

2. Industrial development including the use of technology (such as machines) and the use of energy of sources (such as energy generated from petroleum, nuclear energy, and electricity) for the purpose of enabling the functions of man, manufacturing, transportation, mining, and food processing. Industrial development has led to air, water, and land pollution due to industrial wastes.

3. War results in loss of life and environmental degradation. Weapons used in war can lead to mass destruction. It is estimated that over 100,000 people died as a result of two atomic bombs dropped in 1945 on the Japanese cities of Hiroshima and Nagasaki.

It has been observed that these human activities have affected the atmosphere, causing global warming, changed weather patterns, flooding, diseases such as cancer, famine, pollution, etc.

The following poem is a message dedicated to all humans to support global environmental initiatives and campaigns.

A Cry from Mother Earth

Listen to me, my son
I was a virgin when I got you
I taught you the ways of life
We learnt to live together
Now you are grown
You have learnt your ways
Without my tedious guidance
With power, you control

I taught you the ways to handle
A delicate virgin like me
To keep ice cold
Never melting
To keep waters wet
To keep green alive
Reduce your desert footpaths
For I offered enough, plenty

You failed my lesson
To learn to live
My blood is polluted
East, west, north, south
My brain is melting
My body stinks
I have no air to breath
No fresh leaves to swallow

For this, I will punish
No desert to stand on
No ice to cool, hot
More waters to sink you
More hunger to starve
No air to breath

And diseases to ravage

I give you the last chance
Time's running out for you
You know the right thing to do
To restore my desired state
To restore my virginity
To learn to live together
With your power to control

Forestry and Agriculture

We have already discussed how early agriculture enhanced the chances of human survival and the improvement of technology, including the invention of the wheel. We now explore the role of agricultural development from the rational consciousness stage up to the maturity stage, which facilitated the succession period of the *sapiens* hominin. This period began with the eighteenth century Industrial Revolution in Europe.

During the Industrial Revolution between the eighteenth to the nineteenth century in Europe, North America, and eventually the world, agriculture, manufacturing, mining, transport, and technological development increased the need for timber, and more timber industrial mills emerged. This resulted to further exploitation of existing natural forests as raw materials for construction of houses and industrial use, leading to disproportionate depletion of natural forests followed by deforestation in some parts of the world. A social response to control further depletion of the forest resources then began as realization dawned on the industrial players that the natural resource would be completely exhausted. Such social responses included the Crown Timber Act of

1849,[17] which regulated exploitation of forests in Canada. As the timber resource got scarce, new methods of sustaining its market demands began.

In 1888 in North Carolina, USA, George W. Vanderbilt, the owner of Biltmore Estate, invited Frederick Law Olmsted to oversee the design and construction of the gardens and grounds encompassed by the magnificent estate. Upon Olmstead's recommendation that the estate required a forest manager, Vanderbilt hired a young man by the name of Gifford Pinchot, who developed and implemented a forest management plan for Vanderbilt's forested holdings.[18] Although the rest would be historical, this account is symbolic of any other event unaccounted for that led to the beginning of systematic sustainable use of forest resource as practised today; this symbolizes humans' adaptability to sustain their own survival in the evolution cycle. The study on Biltmore Estate proved that land once badly abused by the exploitation of its forest cover, wildfires, overgrazing, and erosion can be restored and made productive in time through wise management. And this became the basis for afforestation and sustainable land use in the contemporary times.

Domestication of plants and animals marked the beginning of agriculture. Domesticated plants and animals have over the years gained some symbiotic relationship with humans such that their survival has become entirely dependent on the humans and, at the same time, humans have become dependent on them. The human agricultural genius has resulted in genetical changes in domesticated plants and animals over the years by perfect adherence to

---

[17]   *Canada statutes*, 12 vict., c. 30: Southworth and White, 195–209; second report to the select committee on lumber trade.

[18]   http://www.cradleofforestry.com/site/home/history retrieved 11-11-2014.

the *natural selection theory*, where plants and animals with preferred characteristics are selected successively, leading to possibly completely evolved breeds. The animals that are currently domesticated were hitherto wild in the wild with more aggressive traits but are now tamed. Their current organic characteristics have been over the years altered genetically through cross-breeding, location changes, introduction of new feeds, and veterinary medicine among others. These mutations have produced phenotypes that are reliant on humans.

Early domestication of plants and animals was dependent on fertility of the soil and availability of natural pasture for livestock. It was the response to the unpredictability of what the natural environment provided for a community with hunter-gatherer survival behaviour. As a human response to diminishing food from nature, this practice has ensured the survival of the species, and general environmental shocks, such as famine and starvation, have been minimized by ensuring food even during incapacitation and old age, making man a largely agrarian being.

Other innovative responses to enhance productivity to cater to increased human populations since the Industrial Revolution have been:

+ crossbreeding of animals
+ use of manure and fertilizers to enhance fertility of used agricultural lands
+ incorporation of machines such as tractors and other efficient machine-driven farm implements for food production
+ development of the fertilizer industry
+ invention of the mechanical thresher, reaper, and seed drill
+ development of agriculture and forestry research curriculums within the educational systems.

Genetic Modification of Foods

As I mentioned before, selection of the best or more suitable crops and animals over the years have resulted to foods that taste totally different from their ancestors. As a result of biotechnology, these processes have been hastened through genetic engineering to produce genetically modified (GM) crops. Induced mutation to assist breeding has resulted in the introduction of new varieties in rice, barley, apples, and bananas among 3,200 officially released mutant varieties from 214 different plant species in more than 60 countries throughout the world.[19]

Consequently, one can now obtain several varieties of the same plant or animal food, which has a totally different taste depending on its genetic make-up. Of interest is the creativity of man that despite the shrinking space for food production due to population growth and settlements displacing farmlands and forests, there has been enhancement of food productivity. This means the ecosystem bequeathed to the next generation is continually being made artificial to fulfil food requirements.

Many supporters of GM have argued that this is the best technology that can meet the food production requirements of the population and, at the same time, eliminate threats to human survival, such as extreme poverty and food insecurity, while conserving biodiversity.

To the critics of GM technology, I pose a question: what is the other solution to feeding the increasing world population estimated to be expanding to 8 billion on the unchanging quantity of natural resources? Whereas the debate on the potential effects of GM rages with no concrete

---

[19]   http://www-naweb.iaea.org/nafa/pbg/index.html   retrieved 17-11-2014.

proof, the overall outcome is definitely a natural process that should be expected in an evolving ecosystem.

The end result is that GM technology is the *panacea* for human survival to the next level. Just as the historians and archaeologists of this century have stated that the discovery and use of tools were what differentiated the hominins from other animals in terms of ability to survive, the future historians and biologists will credit the survival of humans to their innovations in biotechnology and adoption of the genetic modification technology for food production, whatever the physiological form man will have as a result.

Human Habitat

Human habitats have developed from non-structural habitats to structured and planned habitats. In the beginning, man could have possibly lived like other animals in the forest without any fixed structure dedicated for habitat. Obviously, without tools or with inferior tools, it was not possible to construct a structure for habitation. Perhaps the nearest to what would be equated to the contemporary safe and permanent habitats were caves. Therefore, as tools continued to improve, human habitats started becoming structurally different from those of other animals. But as communal beings they started developing their exclusive habitats in selected kinship groups. Given that man evolved over years being hunter, gatherer, fisher, and then agriculturist, shared habitats provided some security arising from collective response to threats. These earlier habitats or villages could mostly be linked to families, kinships, or tribes and, as I mentioned previously, those who shared hunting culture.

It was not until the eighteenth and nineteenth century, when people of different races and cultures started residing in common habitats or mutually within the same locations that the Industrial Revolution began to change humans into seeing industries as a source of livelihood other than the traditional agrarian economies which were purely based on agriculture. During this period in Europe, trade and commerce grew and thrived in areas where markets were available (i.e. towns, urban areas, and cities), ushering the development of urban habitats for the workforce and traders, who provided service for the industries and supplied goods and services for the urban dwellers.

Although urban areas have emerged for different reasons in different parts of the world, the compelling force for movement and concentration of people within a specific area has always been a similar cause, regardless of whichever part of the world, race, or even class. It is the same one that compelled the basic hunter-gatherer to reside next to the forest with a rich ecosystem—hunting proximity.

Globally, there has been a sharp drift of populations from rural to urban areas over the last century. But this is not a predetermined or planned behaviour; nor is it deliberate. It is part of the processes that fulfil the evolution needs for the new humans, who have continued to be more reliant and dependent on one another for their own survival, and is a reaction to diminishing natural resources.

Urbanization is therefore an adjustment towards an evolution process, and today, more people live in urban areas than in rural areas, with 54 per cent of the world's population residing in urban areas (2014). According to the United Nations report World Urbanization Prospects 2014, in 1950, 30 per cent of the world's population was urban, and by 2050, 66 per cent of the world's population is projected to be urban. In today's increasingly global and interconnected world, over half of the world's population (54 per cent) lives in urban areas, although there are still substantial variability in the levels of urbanization across countries.

How is it a relevant behaviour that contributes to the argument in this discourse? The overall outcome of urbanization is the reorganization of human settlements to enable other parts of the earth to thrive for sustainability of the ecosystem. The characteristics of urban settlements are that more people are now able to reside within a smaller location or area, thus giving rise to housing structures and skyscrapers capable of accommodating more persons within a smaller size of land.

Given this habitat behaviour, man is most likely able to be completely urbanite in the next thirty years, leading to the interconnectivity of the human species from whichever habitat in the world for complete globalization, controlled habitats, enhanced relationships with other animals in the ecosystem, and a global sociocultural behaviour.

In contrast, for the archaic *Homo sapiens* and their predecessors, there was plenty in nature that could sustain them without delineation of areas for food production. Further, the population which settled in a given natural habitat did not adversely affect the ecosystem by the mere act of settlement as the contemporary human.

Education

Education was the beginning of the *rational consciousness stage* of the human species, differentiating them radically from other animals. It is for this reason that the hominins' successive generations carried forward some advantageous social traits and knowledge, leading to consecutive improvements on survival modes. Whether formal or informal, education is both experiential and transmittal through learnt experience of others and even intuitively to affect human behaviour.

There exist different versions of how education has evolved since the beginning of time. The explanation here from the beginning relates to key relevant milestones of education in the human evolution that relates to form rather than substance. In the beginning, prior to *Homo sapiens*, historical accounts indicate humans were characterized by low intuitive and cognitive abilities and limited rational capabilities that are almost similar to any other animal in the jungle, whose motor senses are driven to action by mere needs for individual survival.

In Plato's three aspects of the soul—that is, rationality, will, and appetite—the beginning is characterized by appetitive wisdom. Mostly, the appetitive needs were unrelated to the collective communal objective of continuity of the species. It is an emphasis on individual survival, which coincidentally results to the collective survival of the species. This is the era dominated by human predatory survival culture. As human interactions continued, communal understanding began but only on the basis of the initial experiences of existing threats, food, and reproductive needs of the individual. This relationship is a hunting–protection relationship to derive collective synergy.

With an already developed vocal ability, community groups started developing languages by relating sounds that they produce to existing objects, threats, food, and basic communal interactions, like seduction, tools, etc,. These were carried forward by consecutive generations. This could be discussed further, but it indicates accumulation of knowledge and increased competitive advantage in the animal food chain as successful actions are repeated and perfected. This learning made them prefer some plants over others and animals for hunting or taming for domestication.

Hypothetically, the psychological effect is that the memory glands of the brain adapts to these knowledge accumulations so that the next generation is born already modified to accommodate more experiences through the physiological expansion of the cerebral cortex and hippocampus. The outcome is the enlargement of the brain matter and the size of the head. Consequently, encephalization of the completely encephalized species is supported by the length of human childhood and growth, during which the brain is exposed to some learning experiences, as biologists and psychologists have discussed.

Therefore, the enquiry into the nature of education today is the most germane to determine whether or not these current systems of education as designed is relevant to the evolutionary demands of an ecosystem that prepares for a successive species. Assuming humans have completely encephalized today, with a brain size of 1,330 cubic centimetres, as what the contemporary man has (and this is the same brain size that humans have possessed since the existence of *Homo sapiens* and then *Homo sapiens sapiens*), then there is a possibility that something in the anatomy of the brain has drastically changed. The lifespan experience of a hominin is much more than the experiences of two predecessors within their whole lifetime.

Having agreed that exposure to learning experience, direct or indirect, was the cause of encaphalization, then the nature of brain changes that have happened during the lifetime of *Homo sapiens* in the last 120,000 years must be of a different kind, considering that *Homo sapiens* has had more experience yet the size of the head has remained a constant average. Similarly, going by the theory of differences in lifespan experiences during different stages of evolution of the *Homo sapiens*, it is evident that each experience period has led to different kinds of education systems.

Initially, education was experiential, then it became institutionalized informally, then it became formal for religion, art, philosophy, and then become specialized (e.g. military, artisans) to meet requirements for technological needs. Now it has further developed with the building of academic institutions in dedicated locations that are purely for learning and research.

The modern formal education is now over 200 years, yet there have been a lot of changes in human behaviour and technology. Having attained rational conscience, man has still little control on the evolutionary causes as much as he understands them, making it difficult to be static in the most desirable period and time. Therefore, as these changes begin to occur, motivated by change agents or triggers, how man adapts to enable him to respond accordingly determines his future. The species group that delays to adopt relevant education becomes the one that remains behind in the evolution process.

While it is not a contested fact that some parts of the world had not conceptualized a standard education system as it is universal today, some form of literacy existed within those environments. As it is today, literacy is indicated and measured by how people can read and write. In the case of most parts of internal Africa, and Asia, oral literacy was considered most important. Within the informal structures, grand parents were considered teachers. In the case of my personal experience, it was almost mandatory to sit around my grandmothers cooking firestones where she would give us some lectures about our culture, history, religion, do's and don'ts etc. It is upon satisfactory recital of the previous lessons that she would be satisfied to move to the next lessons. There was no lecture plan or any written curriculum although there was some consistency in chronology of her lectures for each age set of her grand children. Out of these lectures, I am able to recite the names of my ancestors upto my 20th great grandfather,Likung who migrated from from South Sudan to northern Uganda over 600 years ago among other important cultural history imparted to me. In other words the measure for literacy was viewed more in each and every person's ability to memorize what has been taught and experienced. It is this memory levels that determined whether one would be a medicine man, predict rains (sometimes called rain makers), be midwife among other informal professions that existed then.

But oral literacy could not sustain the next evolution level of rational consciousness where so much had to be learnt, creating need for accumulation and storage of more knowledge that would not otherwise be sustained by just oral memorization therefore requiring coding,recording and reference resulting to libraries.

Credit to colonization as one of the ways in which culture is transmitted, the colonised regions began to accept the literary standards of their colonizers which becomes useful in globalization as a cause for re-unification of humanity and non-indifference.

Evidently, over the last few decades, after colonized states acquired their independence, all states adopted aspects of the education systems of their former colonizers. For this reason, various curriculums of independent states were developed in comparison with the education systems of previous colonizers. Having learnt the importance of education in rationalization of cosmology and human relationships, it was now upon the early intellectual to derive relevance from this knowledge to the advancement and decifer its relevance in survival and well being of the new independent states. The critical aspects to advance the cause of evolution which had lagged behind are medicine, governance, philosophy, education, technology and artisanship.

Whereas all scientific research reports on world literacy undertaken by various international organisations such as UN human development report on the education index for 2007–2008 have shown little progress in overall literacy levels in former colonies as opposed to high literacy levels for former colonizers(in the last colonization wave), human beings are moving towards a universal standard of education that suits evolution purpose. Fulfilment of the Millennium Development Goals (MDG) on universal education therefore propels human beings to that next level aspired as there is evidence of actions and plans for attainment of universal literacy all over the world.

Meanwhile the measure of literacy is mostly based on the levels of accreditation within the formal education system. However, with free access to sources of knowledge acquisition, the informal infrastructures of access to education are slowly but surely becoming the major source for literacy. In other words, one no longer needs to go to school to know how to read and write, or even acquire skills for work specialization as it was perceived in the idealistic sense that still propagate the relevance of formal institutions of learning with designated infrastructures such as schools, colleges or even universities. Ultimately, whether accreditation to whichever levels that includes degrees, diplomas and even doctorates occurs, informal attainments of literacy levels and knowledge as in the case today can still find equation to the formal structures of accreditation. Furthermore education is only useful in advancement of human evolution agenda as long as it is directed towards achievement of survival goals.

Let us explore these two scenarios that exist today on objective of education:

A) Two people, *Tom* and *John*, have both undergone medical training. Tom applied for medicine because he wanted to be as rich as the doctor in the neighbourhood, while John was inspired to undertake medicine after he saw the suffering of his grandmother who died of cancer. After completion of college, Tom opens up a clinic and charges medical consultation fees, while John proceeds to research and study more on the causes of cancer. After five years, Tom has become wealthy, while John has completed his research on cancer, publishing his discovery. After some time, John decides to establish a clinic that only specializes on cancer patients. All patients and even Tom's patients are now referred to John's clinic.

B) Jane and Mary were family friends. Their parents were so close that whenever they went on holidays, they drove on the same car. The two found themselves orphaned after a grisly accident that involved their parents. After a road safety enquiry, it was discovered that their parents rolled after their car hit a pothole on the road. Jane and Mary finished college and were now civil engineers. They were both employed as civil engineers in charge of roads in different counties. Learning from the experience of their parents' accident, Jane saved money to buy a better car that cannot roll after hitting potholes as there were many potholes within her jurisdiction. On the other hand, Mary ensured that within her jurisdiction, there were no potholes, and therefore, she bought a car that had the same make and size as her parents'.

We could explore several examples. But example A indicates how the objective of learning can affect its expansion, while B shows how different people utilize their learning experience differently. The two scenarios have resonated globally as the outcome of the modern education, leading to disparities in development and various models of advancement of human beings.

Certainly, with the contemporary developments in technology, urban settlements, international relations, and liberal cultural trends, it has been shown that more learning now takes place out of the formally established learning areas, institutions and schools. The average direct experience with knowledge as opposed to the traditional knowledge mostly acquired through a teacher–student (or pupil) relationship is very high. Knowledge is acquired anywhere (e.g. on roads, televisions, Internet, radio among others).

The delineation of established levels of education associated with proffesional specialization has been relegated to accommodate the increased urge for accreditation thereby reducing in its importance as measure of intellectual ability expected of those who have attained accreditation in the various fields and is hardly evident in performance of tasks in the specific fields that individuals have specialised into, save for a few professional trainings. Furthermore, in most higher learning institutions of the world today, individual intellectual accreditation is guided by the market needs and capitalistic requirements of the academic institutions rather than the general requirements for knowledge as it translates to the evolutionary needs of human beings. It is for this reason that higher learning institutions have become secondary recipient to practical ideas in the market. To explain this further, most recent innovations that have changed the world have had no relationship with the academic nurturing by formal systems offered to their innovators. It is not surprising therefore that most internet hackers have no academic background in computer programming, that a political scientist or even an individual with history as only formal education background would come up with such a software programme that would assist in financial banking solutions etc. Evidently, a vast majority of graduate level and technical college graduates today are not able to utilize the skills acquired in their education experience to support the activities they partake daily for the sustenance of their livelihood in a fast-changing environment. In the long run, the extracurricular knowledge is the one that drives most economies today.

The next education system for hominins is therefore a system which acknowledges pragmatic requirements and is capable of equating extracurricular ability to set standards that can be accredited as the meeting point between formal and informal systems of education to enable adequate monitoring measurement of literacy. Today, a child at a normal kindergarten is capable of logging on to a computer and searching for games that she desires to play, making phone calls, and switching on digital television sets among others since they have acquired knowledge as a result of exposure in their juvenile lives. It is this informal knowledge of the 'new world' that has to be formalized to revolutionalize curriculum development in the education systems.

An attempt to remodel education systems has been done mostly at the most basic level—kindergartens and primary schools. By acknowledging the role of education as opposed to trainings based on targeted jobs only, a system has begun where the school helps in talent identification and nurturing. Kids acquire the basic universal knowledge required for global interaction while pursuing talents identified and their areas of interest without external considerations, such as the market or career development.

Maria Montessori's model is an example of such models that have proved as some of the most useful methods for child development. A child goes to school, and her talents are identified using practical learning experiences. She pursues her talent and develops skills up to her best level. She then uses her knowledge to identify a career that suits her training and talent.

In summary, education has evolved in the following stages since the archaic times:

a) *Conceptualization stage*: This is where the idea of creating structures and institutions assigned for the sole purpose of education-targeted or universal for specific ideological indoctrination such as religion begins. Within the informal set up, cultures begin to recognise that there are a group of members of the society whose knowledge and experience that is transmitted is likely to give competitive advantage for survival if acquired. Schools of medicine, philosophy begin to emerge formally.

b) *Accumilative stage*: after recognition that knowledge can be formalized and transmitted for the benefit of the next generations, consolidation of different knowledge begins. This is the beginning of more writings and creation of ancient libraries that have become reference points on historical events especially in regions where writing was embraced earlier and all over the world today.

c) *Standardization*: Comparative studies of various historical experiences and empirical facts as recorded and observed in various locations lead to realization that knowledge is standard. Each recorded or recited knowledge contributes to meeting an acceptable threshold in education. In the end the realization for standardized knowledge for all human beings is driven by the notion that all human beings need to be equipped with a certain minimum level of knowledge to survive and to achieve their universal rights to equal treatment. This is where we are. The world is in its first step in standardization as revealed by various global efforts such as universal education and deliberate spread of information technology as an emerging source of Knowledge.

d) *Application stage*: Once standardization is complete, knowledge for survival in whichever part of the world will be accessible to all human beings without any variation. At least, in professions such as medicine, there are now globally accepted standard proceedures that are published by the World Health Organisation (WHO) and other global scientific research institutes.

Health Systems for Succession Ecosystem

First, we acknowledge that the health status of any organism is the major determinant of its ability to survive. I will discuss health in the understanding that it is the state of complete physical, mental, and social well-being and not merely the absence of disease or infirmity, as was defined by Henry Sigerest in 1960. It is this definition that fully captures all the aspects for discussion that seeks to clarify that indeed the health adjustments that humans have taken over the years have furthered the evolution cause. To discuss the health of humans holistically, the discussion has been narrowed down to the status and development of medicine over time.

As we all know, humans are at least attacked by one disease or another during their lifetime. Therefore, our ability to manage or treat these diseases have a bearing on our life expectancy and continued performance of social and biological expectations. Ever wondered how an ape or monkey survives in the jungle after attacks from diseases? For apes or monkeys, their bodies are either adapted to carrying severe diseases or they could have strong antibodies that take care of pathogen attacks to the body. The third reason is, they could by chance chew some vegetation, even without knowledge of its medicinal value, and survive. This could also apply to how the earlier hominins could have survived at the archaic stage of the *Homo sapiens* given their scanty medical backward then.

Nevertheless, within the three transiential stages of *Homo sapiens* evolution we are now able to explain the development of health systems and medicine as the *Homo sapiens* progressed through evolution.

At the beginning, after the evolution of the *sapiens*, the archaic group practised magical healing, where treatment was by chance and without the conceptualization of the causes of diseases. Death arising from diseases is justified on the basis of a set of myths that have been created on the basis of symptoms experienced. Then the next stage of rational consciousness results to rational medicine, where some form of aetiology begins from limited the idea that diseases have sources eg rats for plague and not cause such as bacteria or virus and finally came the identification of the relationship between health and diseases. The most prominent record of rational medicine is in the era of the Greek physician Hippocrates between 460–377 BCE.

As humans developed and challenges to survival caused by diseases increased, experimental medicine began. Experimental medicine was further advanced by Robert Koch, who discovered the existence of microorganisms as causing diseases; this resulted to scientific medicine.

With the developments in medicine, there has been steady increase in life expectancy of human beings since the early hominins as a result of some of the listed medical discoveries acknowledged below:

+ Penicillin was discovered in 1928 by Alexander Fleming, resulting to a medicine that could cure a wide range of bacterial diseases, such as pneumonia, sore throat, and septic wounds among others.

+ The discovery of vaccination began with the immunization against smallpox, when Edward Jenner (1749–1823) noticed that cowpox provided immunization against smallpox.

+ The discovery of dental anaesthesia in surgery by William Morton (1819–1868) has helped to ease pain in surgeries.

+ Louis Pasteur (1822–1895) had noticed that bacteria microbes caused diseases and found that heat killed bacteria. This resulted into the development of pasteurization process and its use on food such as milk to prevent bacterial growth. He further developed vaccines against cholera, anthrax, rabies, and smallpox.

+ A German scientist, Wilhelm Rontgen, in 1895 discovered that electromagnetic radiations (X-rays) could penetrate solid substances that light cannot, making it possible to view internal body organs today.

The evolutionary milestones in medicine that completely changed the species include transplant of body organs, such as kidney and heart; innovations on machines that are capable of facilitating or supporting the functions of body organs, such as eye lenses and the kidney dialysis machine; and a further breakthrough by Dr Ian Wilmut and a team of scientists, who successfully cloned a sheep, indicating more possibilities of perfection in medical science that could perhaps lead to the production of human beings in a lab. The outcome of these discoveries is that the species is able to manage the challenges at the next higher level.

Governance

Governance has evolved at the same rate of human evolution to suit the organization needs for each stage of human evolution. The historical period of the archaic *transiential stage* of evolution is characterized by feudalism, then the rational consciousness stage by aristocracy, and finally, the maturity stage (contemporary) largely by democracy.

Feudalism worked well for groups in a largely anarchical world, where feudal chiefs or lords had the organizational ability to protect their respective groups from attacks emanating from other groups. In Africa, feudalism was experienced mostly within kinships that revolved around one within the group with the organizational ability to protect the whole group. This form of governance lived and served its purpose as far as human governance and interaction within the prevailing environment was concerned and, on its expiry, ushered in aristocracy.

Plato and Aristotle, both Greek philosophers, made enquiries that would define aristocracy for long and become the resonance for social contract theories of the nineteenth and twentieth centuries. But the aristocratic period was characterized by both ideas—either philosopher-king or the law. So those with Plato's ideological conviction embraced the philosopher-king concept; a sole leader who has answers to all requirements of his subjects. Therefore, each king expanded his territory for the benefit of their subjects, and each territory became a kingdom where the ruler appointed some of his subjects to help him rule. Nevertheless, there are communities who, unaware of Plato's idea, moved to embrace know-it-all kings or, in the case of Africa, chiefs, an indication that governance was changing globally in its form. The African rulers were expected to be the best medicine men, high priests, best warriors, hunters, and farmers as well as their contemporaries in Europe and Asia.

The Aristotelian rationale for rulership formed the basis for which development of rules and enhanced participation of people in governance is found upon realization of the weaknesses inherent in kings and rulers with permanent roles. After colonization of some parts of the world and their subsequent independence, the people and generally the subjects' awareness of their role in leadership and expectations of leaders increased, leading to selection of rulers through people's participation and democracy thanks to the Athenians.

The American War of Independence (1775–1782) was the midwife to the new world of greater people participation as this was a revolt against aristocratic rule. In this new rule, the people expected a blend of both Plato and Aristotelian ideas of leadership that they also apprise periodically by themselves. Progressively, other states and territories that were aristocratic dictatorships started to adopt democratic systems with the hope to replicate American success story in development, leadership, and stability. Additionally, American engagement with the world in trade, economic initiatives, philanthropy, and international diplomacy was accompanied by American ideological indoctrination and ushered the era of agitation for liberalism and democracy across the world.

With constant changes occurring within the human social structures, forms of leadership must constantly be reviewed. Likewise we must now begin to make a critical assessment as to whether the form of democratic systems as practiced today continue to be relevant to governance needs in this globalized context of governance where the sole role of the state has been majorly relegated to enforcement, monitoring and implementation of social justice needs of citizens.

The initial governance experiences necessitating these changes in developing world was the behaviour of the first democratic leaderships that manipulated the laws and continued in some instances to stay in power illegitimately and therefore not serving the expectations of the people while becoming dictatorial. A part from the dictatorial behaviour learnt in the developing world, past leadership regimes of the developed world had already triggered a dichotomy of Universal expectations on governance and on the other hand individual leadership interests which was larglely mis/interpreted as interest of the states within the principles of self determination. Progressively, the international leadership ideals are beginning to superceed the regional and internal aspiration of governance instruments for citizens making territories mere immigration checkpoints.

As if German philosophers Friedrich Engels and Karl Marx were prophetic in *The Communist Manifesto* on the outcome of class struggle between the working class and the bourgeois capitalists the world is moving towards a classless and a stateless one. Whereas a classless society is in the offing as a result of instruments of social justice currently structured all over the world to guarantee all human beings their rights to be treated equally, the continued territorial management of states has made social justice contextual. In other words social rights vary from one region or state to another.

Justice System- Globalization and Social Justice

After completely exhausting the rationality of laws, exploring the best structures for provision of justice, and implementing theories perceived to result to fairness in the application of laws and maintenance of rule of law, man has still not reached the satiable peak to the justice systems and rule of law. Whereas the justice systems have evolved from non-methodical to a methodical system that derives laws and judgements based on agreed procedure's, the justice system continues to grapple with challenges of erroneous judgments arising from dilemma between rhetorics and scientific facts, idealistic and pragmatic judgments, and global versus local contextualized interpretations of universal rights among other challenges. Consequently, geographical disparities in levels of investigative technology applications, interpretation of empirical facts and contextualized ideological leanings or beliefs have resulted to variant interpretation of laws and thresholds in making decisions on criminal and civil cases against the universal expectations that should ideally defy locational or cultural jurisdictions while delivering justice. The result is that human beings are gradually becoming more dependent on the international structures and systems as they are relatively more standardised as to guarantee justice and equality for all. Amid a changed habitat that is globally interconnected and the universality of some sets of common expectations in the justice systems, local laws of defined regional jurisdictions have begun to cede part of the most fundamental requirements of the populace to a higher and more defined international justice mechanism.

And with the increased interconnectivity of man and the realization of commonness, minimum adherence to some sets of international standards of behaviour is increasingly expected within all jurisdictions. This becomes the basis for existence of institutions such as the International Criminal Court. The trigger to non-indifference of humanity was the Holocaust, which was perpetrated by the Hitler's regime in Nazi Germany (1939–1944) and the effects of the Second World War. The first step was the universal declaration of human rights by the United Nations.

Photo of Andri W. Carnegie- 1835-1919. The man who dedicated his fortune to the construction of the Palace of Peace at the Hague –Netherlands,where the International Court of Justice operates today.

Over years, since the formation of the United Nations and its functional units, there have been joint efforts on achieving minimum requirements for human survival on health, education, human rights, and economic growth among others issues in which all states contribute equitably. Further, many treaties have been signed between states and the United Nations relating to standards agreed upon as *basic minimum human needs* common to all for enforcement and adherence by various actors. Justice is now more inclined towards social justice and social protection which intertwines with economic rights, political rights, and individual rights as part of other inalienable rights to form part of *enforcement obligation* that governments now towards their citizens. The theme of the current non-indifference of the world has completely redefined justice to be consistent with the narrative of realization of mutuality, protective and apathetic relations that make man at par with one another regardless of race, class or economic status, and gender. The guarantee to achievement of all rights by all humans is summed up in all activities undertaken towards social protection today as there is a growing consensus that social justice is the first step towards enjoyment of all the other rights.

Social Protection-Social Justice

Social protection as the most important of the social adjustment programs that are currently undertaken majorly defines the direction of social justice and is generally those efforts taken to ensure all people lead a decent life in order to enjoy all the universal human rights enshrined in the Universal Declaration of Human Rights as adopted on 10th December 1948 by the United Nations General Assembly. Social Protection therefore finds justification within the following articles of the Universal declaration of Human rights:

a) Article 22: Everyone as a member of the society has a right to social security and is entitled to realization through national efforts and international cooperation and in accordance with the organisation and resources of each state, of the economy, social and cultural rights indispensible for his dignity and the free development of his personality.

b) Article 25: (i) Everyone has the right to a standard of living adequate for the health and wellbeing of himself and of his family including food, housing and medical care and necessary social services, and the right to security in the event of unemployment, sickness, disability, widowhood, old age or lack of livelihood in circumstances beyond control; (ii) .... All children, whether born in or out of wedlock, shall enjoy the same social protection

In a contemporary sense, the two articles tackle most of the expectation of citizens towards the states as they exist today. The international non-indifference, global citizenship and brotherhood that defy state jurisdictions is further supported by the following articles of the universal human rights:

a) Article 1: All human beings are born free and equal in dignity and rights. They are endowed with reasons and conscience and should act towards one another in the spirit of brotherhood.

b) Article 2: Everyone is entitled to all the rights and freedoms set forth in this Declaration, without distinction of any kind, such as race, colour, sex, language, religion, political or other opinion, national or social origin, property, birth or other status. Furthermore, no distinction shall be made on the basis of the political, jurisdictional or international status of the country or territory to which a person belongs, whether it be independent, trust, non-self-governing or under any other limitation of sovereignty

c) Article 28: Everyone is entitled to a social and international order in which the rights and freedoms set forth in this Declaration can be fully realized

In fulfilment of the last three articles above i.e articles 1, 2 and 28, there appears to be an unwritten obligation of the developed world towards developing nations that have not been able to meet their obligations under articles; 22,25 and all the other articles of the Universal Human Rights Declaration including amongst the developed states themselves. Both formally and informally, the developed nations have come up with support systems for fulfilment of these declarations through development cooperation's with governments in developing countries, direct funding of the United Nations functional departments which are directly involved in implementing social protection programs such as IDA (International Development Association), UNICEF (United Nations International and Emergency Childrens Fund) and UNDP (United Nations Development Programme), direct funding through their development agencies such as DFID(Department for International Development Assistance-UK), USAID(United States Agency for International Development), JICA(Japan International Cooperation Agency), among others. Informally, private organizations and Non-Governmental organisations have also developed structures that enable them fundraise and facilitate realization of these rights both within the developed states and for the developing countries.

Enforcement of Rule
of Law- New avenues

Enforcement of rule of law occurs at the state and international levels. Within the states, citizens are expected to adhere to the state based social contracts or legislations that govern them while at the international level, states are expected, as recognised individual entities, to perform their international obligations towards their citizens and towards other states. In this regard, the International Court of Justice which was established in June 1945 exists to resolve disputes arising from interpretation on treaties and giving advisory opinion on obligations and responsibilities of the states and state to state disputes in accordance with the International laws.

It was then realised over years after formation of the International court that it was not possible to enforce the rule of law in some exceptional cases where the state mechanisms for enforcements were inadequate,unwilling or even incompetent especially in case of gross violation of human rights by leaders of the states. Arising from ratification of the Rome Statue, in 2002, the International Criminal Court was established at the Hague, Netherlands to deal with most serious crimes of international concern such as crimes against humanity and genocide.

Within the short span of existence of the International Criminal Court a lot has been achieved towards moderation and reduction of outright human rights violations in the world. Triggered by the Yugoslavian crisis and the Rwandan genocide of 1994, the permanent criminal justice system has now expanded in its mandate under the Rome Statute and reformed to achieve its expected outcome.

This global consensus on support of a court with international jurisdiction to deal with some level of crimes in a world that is rapidly globalizing implies that in the end, the state courts will serve as decentralized units for the international-level systems as opposed to what currently prevails. As a matter of fact, the need for an international-level judicial system and oversight arose from the inability of local systems to serve justice. Fortunately, as the court expands, the first steps have been achieved; the arrest warrants such as the ones issued against Sudan president Omar al-Bashir, whether withdrawn or not, and LRA's Kony of northern Uganda have prevailed in the reduction of further conflicts that may not necessarily have arisen directly from those accused. The sunctions of this court is physically limiting to those individuals' movements and interactions while psychologically disturbing. Additionally, the successful conviction of Congolese warlord Thomas Lubanga Dyilo has given an indication that the court has tooth and enhanced its legitimacy somehow.

Certainly, the joint support by critical states of the world has to some extent encouraged submission to the international justice system. Even without formal submission, there is increased acceptance by an overwhelming world population that the international justice system offers the panacea for justice to humans, who have increasingly become international beings, by defying existing geographical jurisdictions.

The most important additional enforcement agency for an international law enforcement jurisdiction is the Interpol, or the international police, whose jurisdiction is global. As most states start realizing that their citizens are not confined within state boundaries as expatriates, tourists, or dual citizens, the rules of engagement change. Laws start to be applied and enforced equally for all in any part of the world. The role of an international police force will therefore be enhanced once state jurisdiction protocols are abolished. While many states are still conservatively holding to the state level determinism —some states have adopted constitutions that automatically domesticated all international treaties that they have ratified as part of their domestic legal instrument making enforcement by a global structure possible.

The other source of enforcement of the rule of law at the state level happens through domestication of the standards arising from the International Standardization Organization (ISO), which is a global mechanism through which minimum standards are formulated for global application. These standards spell out minimum requirements in relation to what is now largely human interactive behaviour. As mentioned earlier, rules arise first out of the need of a hunting group to define their actions and stakes in sharing the kill. Today, man's main hunting grounds revolve around trade, manufacturing, technology, and service provision. This means that once minimums are spelt out, adherence to the required provision of both goods and services standards are expected.

The International Committee on Standards, which is a committee of the ISO, has been able to release minimum standards for economic, social, technological, governance, and environmental requirements with consensus from United Nations, states, and multinational organizations. Surprisingly, these requirements in some instances spell out the applicability of some treaties that are then enforced by various states. Additional enforcement mechanisms for standardization through various national standardization bodies have assisted in the adherence to the set requirements in most states and somehow offer some alternative justice systems.

Case:

In a given country, a multinational has established itself as a brewer and is also an importer of alcoholic beverages. The government has published regulations that prohibit sale of beer containing above 7.5 per cent of alcohol content and other requirements. These requirements are the basic characteristics of a beer that has been determined fit for consumption in that country. One day, there was a lack of one of the basic ingredients due to delay by external supply sources. The company decided to proceed without the ingredient. The result was that the overall product had 10 per cent alcohol content, which was above the requirements for consumption. But the multinational proceeded to bottle and sell the product locally.

The justification of this act was that the company employed many people and was likely to get losses that would result to job losses and that the revenue generated by the government could be affected if the company did not sell.

So when the product went to the market, it was found to be harmful to the consumers. Fifty people died and another twenty-eight were rendered permanently blind. The directors of the multinational were taken to court and jailed for fifteen years according to the country's laws.

I presented this scenario to a female victim of a related case who had lost her husband in my country and asked if she felt that justice had been done. She replied that justice is not punishment. It is that which could have prevented the deaths since death is irredeemable. I therefore proceeded to ask what she could have felt was just in such a circumstance. She informed me that justice is about each and everyone doing their work or undertaking their responsibilities to ensure all humans enjoy their rights in compliance with the set or agreed standards. If it is 7.5 per cent, stick to it, and let someone responsible confirm that you have not gone beyond.

Most of the activities and expectations of humans towards others are not legislated but regulated by some internationally agreed standards. For example, there is an expected standard length of beds in any restaurants all over the world. That is six feet. The width varies from three feet, four feet, five feet, six feet, etc. So one does not expect to find the bed length requirement in any legislation nor does one expect less. A kilo of any product would be the same weight anywhere or a kilometre is the same all over the world. There is no argument.

Therefore, as we move towards an empirical world, legal arguments are lessened by the set standards which are more likely to be agreed upon at the international level and automated.

The (criminal) case below is one of the many examples where weaknesses of the justice systems at the state level has been exhibited:

Case:

In 1996, a Chinese teenager was convicted of rape and murder. After undergoing a rigorous court process, where all facts were presented in accordance with the modern judicial process, he was found guilty by the court, which ruled, according to the law, that he be executed.

Hugjiltu Qoysiletu was put to death in 1996 in Mongolia.

In 2005, a man confessed to have committed the same crime, and it was indeed proved that the confession was factual. Eighteen years later, on Monday, 15 December 2014, the Inner Mongolia Higher People's Court found that the court verdict in 1996 leading to execution of Hugjiltu was not consistent with the facts presented and that there was no sufficient evidence to make such a ruling.

Even after that ruling, not any court nor the Chinese state could resurrect Hugjiltu. But the dead teenager's mother and father, who suffered the unjust loss, could only burn a copy of the court decision on his grave as a symbol of abhorrence for the judicial system.

Hugjiltu's case is representative of many decisions arising from a judicial system that has outlived its time.

With the limitation of state laws as a means to achieve justice within the globalized world, mans aspirations for justice has moved to a higher level for institutionalization of justice globally.

Religion

There are several versions on how the idea of religion could have begun. However, all the causes and developments in religion can be tracked with the evolution of man over the historical period of the *Homo sapiens* history. These historical periods coincide with the cycle of religiosity, beginning from atheism, then polytheism, and monotheism. The coincidence is such that atheism is predominant at the archaic transiential stage, polytheism during the beginning of the rational conscious stage, and finally, monotheism at the maturity stage of evolution of *Homo sapiens*. Certainly, there is a relationship between a complete evolution cycle of religion within the three stages and evolution of man. A complete cycle of the three religious forms symbolizes a complete cycle of evolutionary stages within the intratransiential stage of evolution. The maturity stage of evolution is characterized by a developed ecosystem that facilitates succession into the initial stage that would be substantially different from the first form in the initial cycle.

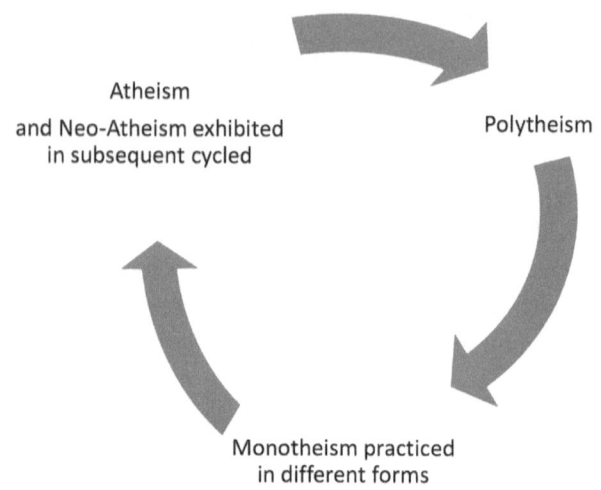

Religion evolution cycle.

Atheism

In a broader definition, atheism is the lack of belief in a deity, god, or any metaphysical idea that forms a religious belief system. The absence of belief can either be an outcome of diminutive consciousness, lack of a conceptive idea for a deity in the first place, or an advanced empirical culture that requires tangible proof to support a belief system in an environment where there is lack of evidence to support a religious belief system. The latter occurs in highly developed communities with scientific knowledge or mastery of their environment.

In the beginning, the interaction of man with the natural environment preconditioned his social behaviour towards fending for food, nurturing, and procreation. Social norms were intertwined with biological behaviour such that there was little room for abstract thinking rather than instinctive rationality that was only limited to survival. At this stage, conceptualization of religion was therefore null, creating a perfect atheistic state.

As the social units strengthened and humans settled, they started developing rules that guided their existence within their various habitats where life experience begun to form knowledge that was memorized (in some instances, stored on stone tablets). Uncertainties and objects that man could not understand or explore in each environment were mystified. These mystifications, organized and transmitted through generations in social units, were eventually transformed into symbolism that changed form to become ideas for which a belief system is organized. It is this therefore that explains clearly why belief systems in the written history of mankind differ from one region to another. It is these outcomes of experiential differences that resulted to a global system of many different deities, ushering *polytheism*.

Polytheism

As each physical form is changed into an idea, it forms the basis to which different natural causes are attributed. For example, the sun, being the giver of light, becomes a deity; the soil, being the base for fertility of land, becomes a deity, etc. The beginning of these imaginations is the beginning of the rational consciousness stage of evolution. In the prelude, all groups uniquely develop their own metaphysical conceptions based on their strange experiences of their own environment. The idea that manifests itself as more relevant to existence therefore becomes the most superior of the deities.

This stage also begins to diminish as rational consciousness stage of evolution grows in age, leading to more interactions between different belief systems that lead to reconciliation of beliefs that have logical convergence. Additionally, as law enforcement and governance structures begin to impact humans, a great shift from worship of multiple deities to unification of belief emerge not only for religious purposes but also to enforce loyalty in governance. This leads significantly to the progressive disposal of the polytheistic religious system.

Monotheism

Submission and loyalty was enforced by imparting a religious belief that was accepted by the ruler on the subjects as a show of subjugation. There is a clear relationship between monarchy and monotheism in terms of symbolisms, architecture, and aspirations. In a monarchy, sovereignty is embodied in one individual who was traditionally the lawgiver and has exercise of absolute powers. Within a territory, it is the ruler's religious belief or most favourable deity that would therefore be transformed into a single that is the most relevant of the deities. As the other symbols are ignored or rendered irrelevant, monotheism begins.

In Rome during Emperor Constantine's rule from AD 311, there were different deities. Emperor Constantine's mother, Helena, being Christian, influenced the ruler who would trigger the demise of other Roman gods and lead Rome to the religious belief of one god. The god of Christianity has the cross as an emblem of worship. According to other historians, monotheism was the ideological tool for which Emperor Constantine consolidated his rule, especially as his rule opened a significant chapter for the reunification of Rome.

This belief in one deity created some order in governance, which enabled indoctrination, leading to loyalty to authority until the scientific renaissance in Europe.

As tool for governance, monotheism (not absolute) as experienced in different environments was sustained even after the separation of State from religion. The various religious groups, mostly Christian, supported ideological indoctrination for legitimizing governance and subjugation during colonization of Africa. Thanks to the Protestant movement begun by Martin Luther in 1517, thereby the time of the *last colonization*, the church was reformed with more liberal thinking, which was mostly from the protestant groups; these complemented the colonialists through education and training for governance.

Development of monotheism in Africa

In Africa, the history of monotheism can be traced to the beginning of colonization. Collectively, each community residing within the different geographical location as kin developed their own belief systems, which could in some cases be different within the same ethnic community. In summary, it implies there were possibly hundreds of deities in Africa.

The approach of the colonialists was a coercive enforcement of their religious belief systems through either Christianization or in Muslim states—Arabization. This forced belief system did not yield much as the African mindset was so entrenched into their traditional belief system. Among the Christians, there was a compromise where culture and religion were separated. The hitherto traditional religious belief system was celebrated as part of culture and daily routine, while the Christian religious requirement of attending churches was obeyed on the designated days of worship.

In response to indoctrination of the monotheistic religions culture, which was spread derogatorily against the existing African culture and as part of the colonization menu, a new set of religious system was developed as a protest to the colonizers' maltreatments, or in some cases, there was movement to other optional religious groups, such as Islam. Most sovereigns in communities, having conceded to monotheism, blended their monotheistic experiences with their ways of life, which was utterly unacceptable to the colonizers' religious practitioners. This was the foundation of the African independent churches.

The case of Nomiya Church in Kenya

Nomiya Luo Mission is one of the early religious groups that have a unique blend of religious doctrines. Started in 1907 by John Otieno (Owalo), son of Abor from the Kochieng clan in Asembo, Siaya County of Kenya (previously, Nyanza Province).

John Otieno's journey began when a British administrator, *Mr Morson* (as he is recorded), took him to work as a cook in the late 1800, where he learnt to read and write, thus becoming one of the first interpreters for the Scottish mission in Kenya. By early 1900, another British administrator, *Mr Ardhwen* (as he is recorded), posted him to the Thogoto mission (Kikuyu tribal area). He became in charge of supporting African children to attend schools, forming part of the pioneer African teachers in Kenya. He taught one of the early renowned independence leader in Africa and the first president of Kenya, Mzee Jomo Kenyatta.

But John Otieno (later known as Johanna Owalo) was not a conformist. During his lessons offered to African students, he rejected some of the most fundamental ideologies of the Christian missions—the trinity and also of Jesus Christ being the saviour of mankind. He taught his students that if Jesus was a Jew and he was sent for the Jews and if Mohammed was an Arab and was sent for the Arabs, then in the same way God would only send an African for Africans. This was logical since during that time, it was a requirement for Catholics to learn some Latin to qualify for baptism and to learn some Arabic to be a Muslim, yet this rebellion triggered ideas that would later form the basis of the independence struggle among his students. When the British area administrator heard of Otieno's ideas, he was transferred to Mombasa as a cook for another administrator. While in Mombasa, he converted from Christianity to Islam. As a Muslim, he changed his name from John to Johanna.

Having undergone religious conversion from African traditional religion then Christianity and finally Islam, he introduced an idea that blended both Christianity, Islam, and the Luo traditional practices, and he called his religion Nomiya, which was earlier reported as 'the mission I was given' in translation. This looked sacrilegious and most controversial at the time. As punishment, he was sent back to his ancestral home in Asembo, where he continued spreading his doctrine among his Luo community. In summary, the doctrine is a blend of religious beliefs that lays more emphasis on one god by accepting traditional religious practices. Nomiya Church practices are 40 per cent Islam, 30 per cent Christianity, and 30 per cent traditional Luo.

Neo-Atheism

The new form of atheism is ushered in by a highly secularized religious behaviour of man, knowing it is possible to blend religion and the normal ways of life. Having institutionalized capitalism as a way of life, religion is now purely influenced by capitalistic behaviour to the extent that it becomes an enterprise. When there is competition rather than complementariness between religious and individual enterprises, then conflict of loyalty arises between individuals and the religious institutions in a manner that affect their faith in such religious institutions (e.g. if the religious group I belong to is my competitor in business, then my faith diminishes and the morale to pay tithe to sustain it diminishes as it gains a secular life).

Secondly, the daily bombardment of knowledge that leads to rational and metaphysical thinking, which disapproves or falsifies some of the religious beliefs, has become a factor for increased atheistic tendencies or rejection of belief systems that lead to religiosity. As scientific discoveries continue to thrive, man's belief in religion declines.

This begun with the writings of *Galileo Galilei*,[20] which had empirical facts that disapproved the prevailing religious doctrines about the universe, leading to his persecution. Since then and after the scientific renascence, different perspectives on religion have been explored, and a further blow to religion was the first arrival of man to the moon.

Thirdly, the organic theory of evolution became more convincing as a rational explanation to existence of plants and animals. Indeed all these reasons among others are the reasons why monotheism has almost completed its period of relevance. And all these relate to the completeness of the cycle of evolution of the *sapiens*. It will not therefore be surprising for the future generations to label religion as a psychological adaptation that enabled the current generation as a blockade to the rationality they could not bear and as a solace for uncertain and uncontrolled occurrences. The future exploration and discoveries in science and technology will virtually give all the answers to the metaphysical attribution of today.

The place of radical religious groups in the future

Whereas the religiosity demographics of the world may indicate an increase in the population of some radical religious groups across the world by 1994, this increase can be attributed to other factors that are non-religious but practised by the groups and other political realignment in

---

[20] The *Dialogue Concerning the Two Chief World Systems* written by Galileo Galilei in 1632 is reported to have been initially titled as *Dialogue of the Tide*.

addition to polygamy. However, it is clear that radicalism threatens the religions of the world today where there is association with violation of human rights, segregation, war, terrorism, and other vices. Man has evolved from predatory and cannibalistic tendencies to mutualism, making vices committed in the name of religion counterproductive to the development of such religions.

These radical behaviours, such as terrorism, religious wars, and other vices, that are undertaken by the conservative adherents with predatory mindsets are repulsive to the growth of religion and help to further rationalize the justifications available for atheism. In a world that is continually becoming liberal, religious beliefs must be persuasively inculcated as religion becomes a matter of choice rather than a reason for living.

Thus, continued association of identification with social ills and loss of human life due to beliefs by the renowned large religious organizations, which have historically been seen as carrying the banner of religion, may lead to an abrupt change in religious behaviour of the contemporary man who has defied religion and culture in his/her interaction. As social labelling continues, and liberalism, which is more fashionable as it guarantees more freedom, becomes the global norm; negative religious behaviour has become the midwife to neo-atheism.

As opposed to other non-religious social organizations, where it is possible to separate way of life from the organization's cultural belief system, the tendency of religious beliefs to encroach on private ways of life increasingly becomes unbearable in the liberal world. It is rare to find atrocities in the contemporary world (by any organized group or state) done in the name of a liberal social movement. Relegions that threaten the existence of other people will therefore naturally be the first religious organisations to exit the human organisation system.

CHAPTER 5

OTHER EVOLUTION CAUSES

Roles Redistribution by Gender

Over the last few decades, different social and cultural definitions have been assigned to the term *gender*. Undoubtedly, all languages or meaning of words are socially and culturally constructed since the evolution of languages. In this case, these redefinitions have deliberately been intended to socially rationalize redistribution of roles between males and females. According to the World Health Organization, *gender* is the result of socially constructed ideas about the behaviour, actions, and roles that a given society considers appropriate for men and women.[21] The rationale for these definitions is limited to the development of a cultural dictate that enhances equality and, in most instances, equity in assignment of roles between a set of sexually different individuals in the society.

The contradiction in this case is that sex is not a social determinant but a biological result. Man does not become

---

[21] http://www.who.int/gender/whatisgender/en/

a woman or vice versa by social construction. *Masculinity or femininity*is a linguistic definition given to the inherent biological characteristics in either the male or female of the human species. These biological characteristics predispose some adaptive traits which are socially adopted for the species to survive and competitively perform the natural roles assigned according to their physiology. Hence, social construction of gender roles including assignments or reassignments of roles is a mandatory command of biological disposition in humans even in the without socially constructed definitions.

The Lion's Pride

A pride usually comprises about five to six adult females, a set or coalition of adult males, and their cubs. Within a pride, males are usually able to scrounge food from the females, but they also have pride duties. Males have to patrol and mark their territory by spraying urine, rubbing secretions of glands on objects, and roaring for security of the pride. Females also mark and roar, and both males and females have to chase or fight off intruders, risking death or disability. Males only defend against other males, while females defend against other females as well as strange males.[22] The lionesses does most of the hunting for the pride, while the male lion stays and watches his young, waiting for the lionesses to return from the hunt, with the kill. Usially, during a communial kill, and even in normal circumstances, the lioness will always feed last after the cubs. In the feeding order, the lion begins then the cubs and finally the lioness remains with the leftovers.

While making this observation at the Maasai Mara, Jeramiah Olepunyua, a guide urges that in this lion protocol of feeding the *male lion* makes the kill softer for the cubs to

---

[22] http://www.thebigcats.com/lion/lion_social.htm.

consume the hard game meet. The lion is ofter the strongest and therefore more able to tear the hard game meat apart.

Lessons:

From the lion's case, we can conclude that the biological abilities of all animals will naturally dictate their social functions due to the natural attributes, which inherently becomes *masculinity* or *femininity* or in whichever terms as socially classified.

Assuming we were to reassign *unnaturally* that the lioness does the roles of the lion and vice versa, then the following are likely to happen to the lion species (considering the physiological characteristics of both):

1. His ability to protect the pride would diminish, leading to attacks on his cubs from other predators.
2. There would be food shortage in the pride because the lion is not as swift as the lionesses.
3. The lionesses would not be able to keep their pride due to exposure to other males.
4. The cubs will not be able to consume as much food required for their development since that ability is only inherent in the 'masculinity' of the lion
5. Due to the inability of the lion to make frequent food supply as required, coupled with the exposure of the cubs from more dangers, reduced protection of the cubs and scarcity of food may lead to extinction of the species.

This leads us to the next question: What is this unique thing in loins and lionesses that makes them functionally different within their social set-up? To this the answer has more to their masculine or feminine biological characteristics that each one finds his or her self possessing

Femininity is socially constructed but is made up of both socially defined and biologically created factors.[23] This applies to masculinity as well. Biological beings must commune first for social definitions to arise. Therefore, we must all agree that the biologically created characteristics of lions determine their sex roles.

Let us assume a second scenario: human beings.

Like the lion and lioness, the human male and female have different physiological attributes. Even humans who have exhibited the *third gender* can only assume either masculinity or femininity. One biological attribute must be dominant while the other is recessive at a time. Whenever masculine characteristics are displayed, it means there is a dominance of the male biological characteristics; femininity also applies to females with the same rule. We can therefore only classify humans as either masculine or feminine regardless of other physiological aesthetics, transformation or otherwise.

The roles assigned to them are based on social rationalization of dominant biological traits that are already classified. These roles may differ or be similar, depending on the existing environment. For example, while women in location *x* are culturally assigned the role of building houses, women in location *p* are not; men in location *y* are culturally assigned the same roles, while men in location *f* perform the tasks of women in location *x*; etc. To assess the similarity of these roles, we must access the magnitude of work and the risk level of the task performed within the variant environmental conditions. Highly risky endeavours are necessarily tasked to the male or one whose loss could easily be compensated for the continuity of the community.

23   Hale Martin, Stephen Edward Finn (2010), *Masculinity and Femininity in the MMPI-2 and MMPI-A* (University of Minnesota Press) p. 310, ISBN 0-8166-2445-3 retrieved 3 June 2011.

The magnitude of work is also assigned depending on the biological endowment of either the male or the female. Sharing of roles occurs only with regard to those secondary functions that can be executed by either, and does not fall within the two justifications nor requires sexual characteristics.

As for roles that are socially constructed as opposed to ones that are biologically given, such roles are definitely secondary, not basic. They are secondary in the sense that they support the basic biological functions that are important for survival and development of the species. Secondary roles do not necessarily require biological attributes in some instances but may have an effect on the feminine or masculine attributes.

The traditional masculinity and femininity traits that determine the quality of secondary roles have evolved over the years as the basic gender roles created biologically remain constant. Naturally, the biological predisposition of performing certain roles could have resulted to adjustments in capabilities by either male or female within a given environment. In a less complex hunter-gatherer environment, each sexual trait helped humans to attain the means for survival, and they were therefore socially assigned appropriately. The male was gigantic, strong, faster, and forceful to enable them to effectively hunt and protect their families, while the feminine attributes were and still are such that attract men for procreation and nurturing of the children. In such an environment, the female is not subjected to an environmental frontier with higher risks as a cultural behaviour that supports survival of the species since she is able to propagate the species with any other male, or in cases where risky endeavors result to loss of men, at least one or two men against more women could still save the species. These traditional roles assigned to either male or females weighted against their magnitude indicates the high level importance attached to the female of the human species over years as a cultural strategy for survival.

Risk levels, sexual role assignments, and complementariness

The normal male sperm count in a single ejaculation is approximated to be at least 280 million sperms.[24] Out of the 280 million sperms, only one fertilizes the ovary for the conception of a human being to occur. Additionally, the male sexual organ at a youthful age is responsive to stimulation every day in a year at a frequency that can only be determined by the presence of a stimulus. In other words, erection of the penis occurs any time it is stimulated, and during each erection, the male is ready for recreational or procreation mating. This implies that to a minimum, one man can cause at least thirty conceptions or more per month. On the other hand, a woman's monthly fertility circle provides for only one chance of conception per month, after which it would take at least eleven to twelve months for the next conception. This could mean that for the human species to survive, not so many men are required after all. On the other hand if there were fewer females than males, that would be a clear indication that the species is diminishing since the rate of reproduction would be lower in addition to childhood vulnerabilities that lead to mortality.

Therefore, over the years, society has assigned males and females social roles that reduce vulnerability, depending on the existing populations in each environment. With full knowledge on the importance of women in the continuation of the human society, the roles assigned have always had lower risk levels, whether cumbersome or light, while roles assigned

---

[24] http://www2.oakland.edu/biology/lindemann/spermfacts.htm retrieved 11-11-2014.

to men have had higher risk levels, whether heavy or light. It is for this reason that in most or almost all communities of the world, men are the ones who go to war or hunted. Thus the measure of success from any war in summary is the extent to which women and children were protected.

Now that it is clear that basic gender roles cannot be assigned socially, we must agree that even before technological development and the long historical struggle for women involvement in development, women are still performing both primary and secondary roles, depending on the risk levels of the secondary role. The reason for disparity in roles across the board that has been confusing in the gender discourse is the risk factors, which varied from one environment to another and therefore in the difference of assigned roles. In the evolution sense, therefore, there have never been any role so discriminately assigned as to favour the female over the male and vice versa. To deliberately reassign roles without cross reference to the natural process would consequently be retrogressive to the entire human species, whose survival is based on mutuality.

My Trip from Mtwapa to Kilifi

On 25 November 2014, I decided to observe the economic activities in Mombasa and coastal region of Kenya and the roles played by males and females. I have been informed that the Mtwapa area is in business twenty-four hours daily. So I boarded a *matatu* at the NSSF stage in Mombasa City.

On our way, we made the first stop at Bombolulu, where there were a lot of activities. It was around 7.30 a.m. People were approaching the *matatu* in pairs of young men (youths) carrying luggage or trade ware and of women approximately in their forties. However, it was the women who were finally boarding the matatu. As we arrived at the Mtwapa market, majority of those in the open market were women, while the men around had *mkokoteni* (hand carts) for carrying loads to the women's market.

I therefore concluded rightfully that the role of men here was to assist women to carry their trade wares after an interesting discussion with one of the women seated next to me, who confirmed this conclusion. I proceeded to my business in Kilifi, where I observed even more-critical roles of women as most of the men's work could not generally be defined apart from the ones in employment. On making further enquiries from Ms Halima, who ran a kiosk in Kilifi, she complained that her husband was a drunkard and contributed nothing to the family save for the land where she had built her ancestral home. I concluded my business for the day.

On my way back to Mombasa, I alighted at Mtwapa at around 9 p.m. to meet a colleague who had promised to buy me a drink. But at this hour, Mtwapa had completely changed as majority of the population around the market and especially the pubs is comprised of young women from ages 18 to late twenties. They were in business. The population of men is also sizeable enough; however, these men were of different ages, and they looked more refined and neatly dressed as compared to the ones I saw in the morning. They had come for leisure.

After one hour, around 10 p.m., my friend Kamau decides to call one of the ladies to join us. This one had sat alone next to the counter of the pub. Her name was Fatuma. So I decided to ask her what she was doing in the pub, and she retorted in Swahili, 'Mimi Niko Kazini!' I am at work. At that point, I could not interrogate her further since I already knew what she meant. My assumption was not wrong. During our discussion, she informed us that she had left her husband at home with her child and that she was the breadwinner of the family. Before we left at 2 a.m., my friend offered her 1,000 Kenya Shillings as a token of appreciation.

She was appreciative but still offered to entertain my friend for the generous gesture. We declined and drove back to Mombasa.

Implications of the Growing Minimum Mutuality in Role Relationships as a Result of Role Redistribution

Minimum mutuality of roles arises where one gender can perform both primary and secondary roles with minimum assistance or in absence of the other. As men and women develop non-dependent sexual relationships, the consecutive generations respond both socially and biologically to the lessened mutuality behaviour. As mentioned earlier, this social behaviour results to such biological adaptation that arises out of nurturing children, which takes longer periods. So, with technological development where all work related responsibilities have been made lighter, there is a diminished role of the male as most of their roles that were naturally assigned out of their generally inherent biological characteristics can now reassigned to the female counterpart with the aid of technology.

Further, the reproductive role may diminish too due to the advancement in reproductive health and other health innovations on reproduction such as the existing possibilities that females can now conceive through artificial insemination. Should all women decide to have babies from sperm banks and conceive through artificial insemination, can only one man undertake the reproductive role of all men on earth today? The answer is yes. However, in conceding to this empirical fact, I would either be contradicting the argument that coexistence and mutuality is the basis for survival or yielding to the contrary argument that there is a possibility for humans to evolve towards extinction as this possibility has been compensated by other factors so that men could still have relevance.

Diminishing Role
of the Marriage Institution

As the roles of males and females converge, the social institutions of marriage is either collapsing or changing form. Apart from nurturing and reproduction of the species, why should the same man and the same woman stay together? Are not these traditional family roles *outsourceable?*

Let's take the example of nurturing. There are now more trained childcare professionals or nannies who take up the roles that mothers used to have in care and nurturing of infants. After the first three years, children are subjected to day care centres, where they play with other kids and learn. Their interactions with parents become minimal as they arrive home exhausted and sleepy. And the routine continues. Later on, the children attend the basic education system, in which the role of parents or guardians is diminished to financial contribution and hosting of the child, who gains independence on completion of the schooling system.

On the reproductive roles of women and men, science has now made it possible to combine sperms and ovaries of two unknown males and females. Vitro fertilization is where an embryo is created using male sperms and female eggs and fertilizing it in a laboratory. As a result of this fertilization, the embryo can be transferred to a totally different female uterus for the development up to the birth of a child. The woman whose womb is used becomes a gestational mother.

These are just but a few examples. Although same-sex relationships could have existed before for a different purpose, the reduced sexual mutuality and available options for fulfilment of the mutuality purposes have resulted to an increase in the number of people engaged in same-sex relationships and transsexual sex reassignments. These numbers are projected to increase as man continues to reassign socially or offer alternatives to the biological functions of both sexes.

Further, there is a general global increasing trend in children born out of wedlock and children being raised by single parents, which is part of the indicators to the above argument. In the United States of America alone, for example, couples between the age brackets of 20–24 are seen to go in for the maximum number of divorces, with 36.6 per cent of women wanting to end their marriage and 38.8 per cent of men wanting to end theirs (US Census Bureau, 2009). The causes and reasons for these divorces are contradictory to the real reasons for a marriage in relation to human survival.

These reasons, to mention but a few ranges from; infidelity, unemployment, financial goals, lack of communication, intellectual incompatibility, etc. They indicate phenomenon changes in human perceptions that could also be attributed to changes in the brain status of the contemporary man who, having mastered himself, his environment, and his purpose, is capable of interrogating his own social action and responding communally or individually to what is most appropriate to his survival. With the possibility of interchanging roles, technology that has had impact on reproductive health and machines that can be operated by either masculine or feminine characteristics, the purpose for the enforcement of social units comprised of two people of opposite sexes is diminished.

The natural reason for the female behaviour that attracts males has also changed from the needs of the ovulation cycle to day-to-day recreational objective. The inherent characteristics that attracted the female to a male have changed from visible to imagined as the most necessary object for attraction to the male is now concealed by clothing, plastic behaviour and artificial fragrant. Of the two, i.e, males and females, the most endangered by these developments is the male.

Response to Diminishing Sexual Mutuality

That is why the world has responded *though unconsciously* to the threats to coexistence between males and females by enhancing both reproductive and recreational urges of the female sex to require the male. One such action is the legislation and cultural education on abandonment of female genital mutilation (FGM). By February 2014, it was estimated that more than 125 million girls and women had been cut in the 29 countries in Africa and Middle East, where FGM is concentrated.[25] In December 2012, the UN General Assembly adopted a resolution on the elimination of female genital mutilation. The overall outcome of this campaign, in my opinion, is that majority of the women, if not all, will have increased female emotional needs occasioned by levels of female confidence, which is likely to result to more emotional needs.

As a consequence of the total abolishment of FGM, and a generally increased level of self confidence in women, recreational sex is most likely to be higher. Recreational sex results from the (likelihood of) increased erotic stimulation in the nervous system of females, necessitating the need for sex. Elimination of the FGM practice will therefore sustain the relevance of males towards females and vice versa in the next generation. In the end, having the knowledge of that, the global efforts towards the elimination of this practice should focus on the beneficiary, the males as a practice learned from the market forces where products are always adjusted to suit the consumers. In this case, the practice will most likely

---

[25] http://www.who.int/mediacentre/factsheets/fs241/en/ retrieved 17-11-2014.

end when it becomes clearly unfashionable amongst men to sexually engage women who have undergone the cut.

It is, however, unfortunate that as the elimination of FGM continues, the world has turned a blind eye to male circumcision, which has almost to some extent the same sexual effects in males as does FGM in females. When the *sheath* is removed from the penis, the sensitivity of the *glans* reduces over time as the erotic nerves in the glans are exposed all the time, reducing the recreational satisfaction of sex by the male. This reduced intensity of sexual satisfaction occasioned by circumcision also reduces male urge for females thus affecting sex mutuality levels. It has been historically proven that traditional societies where circumcision was practised on men had lesser regard for women than the ones that did not circumcise.

Human Nature-Selfishness and Greed

Selfishness and greed as characteristics of human beings are a major contributor to mans efforts for mass accumulation of resources at the cost of depriving of others, leading to poverty, social inequality, and class distribution by humans.

Mass accumulation of resources is characterized by piling up or taking possession of what is beyond the needs of the accumulator. It results in storage, non-use, or misuse.. Wealth results from the economic value of resource that has been accumulated. But the cause for accumulation of resources initially was and continues to be uncertainty of the future or our inability to predict its outcome. In the beginning, whether in Africa, Europe, or Asia, the hunter communities developed a storage mechanism for food through drying or freezing so that should a kill not be there the next day, the family would still survive.

Predatory behaviour include corruption, war over resources experienced between states, theft, and other vices that have been used to accumulate riches at the expense of others.

The initial deprivation of land to support livelihood can be traced as the beginning of poverty and inequalities that exist in the world today and the Marxist class divisions. Thus the traditional reasons for conquests of territories by the pre-modern governance organizations were to increase land space for creation of more wealth by the loyal subjects. Social stratification arising from mass accumulation of natural resources, especially land, which determines the levels of production of wealth by individuals, have over centuries of rational consciousness in Europe and Asia led to classification of levels of importance in the treatment of humans by humans and the reduction of the development equilibrium required for stage-to-stage evolution by all humanity. In Africa, colonization leading to racial consciousness and attributed social stratification based on race further reduced interactions and relationships.

Let's take Asia for instance—the caste system. This system, which is based on wealth, prohibits or limits sexual relationships between different castes, leading to inbreeding within the same caste. Consequently, a generation with genetic deformities or with the inability to withstand environmental challenges gets extinct, leading to the disappearance of some of their gene traits that would ordinarily form part of the available global pool for enriching the genetic hybridization of the human species

Most wars wedged and fought were for territorial (land) claim that is related to other resources. Even the 1990 2 August 1990 Kuwait invasion and territorial claim by Iraq, which led to the Gulf War of 1991, was based on land with oil resources.

The initial cause for the *great human migration* was greed and selfishness, leading to territorialism, wars, and subsequent displacement of some groups from a habitat that the hominins had lived in for millions of years. My imagination of this event that happened approximately 150,000 years ago is that if whatever happened then is replayed today, it could cause a *global humanitarian crisis*. You can imagine a group of humans migrating for safety and food as a result of displacement across the semi-arid regions of northern Africa (*the great migratory corridor*), then across the Mediterranean Sea, and to a winter in Europe, which they have never experienced before! It was definitely a horrible experience for the first generation of the *sapiens* to settle in Europe.

All wars, displacements, and human suffering have come as a result of greed and selfishness, which have, as negative as they are, had a share of contribution in the transformation of the hominins.

Human Nature-Discrimination (Racial, Religious, or of Any Kind)

I can't breathe!

The phrase '*I can't breathe*' will haunt the United States of America for years to come. It was the last cry by Eric Garner, 43, from Staten Island in July 2014. He was allegedly selling loose cigarettes on a sidewalk when he was put into a chokehold by a police officer, Daniel Pantaleo, 29. The video capture of how this incident occurred was spread across the world on YouTube, Facebook, and all *social media*. It depicts a behaviour similar to that of a group of hyenas attacking and devouring an elephant. This unarmed gentleman was attacked and put on the ground by around four officers, with Pantaleo holding his neck. He cried out that he could not breathe until he became motionless. The unfortunate part of it is the mob psychological behaviour of the officers who were not only unable to restrain the one holding Eric's neck but also continued to pin him down as he died. What is conspicuous about this scene is that all the officers have racial difference from Garner.

Eric Garner's death remains a symbol of acts arising from discriminatory inhuman treatment by others of different *varieties, class, race, religion,* and *other social divisions* created through socialization where the mindset of the offender is that the others are lesser beings. In the case of Eric, the consequent reaction of the offenders happened with a person of a different *race*. So we want to tackle racism. In the absence of racism, there is also class, tribe, and religious divisions that result to the same outcomes that are divisive and dehumanizing, depending on socialization.

Extreme behaviours have always segregated humans and led to discrimination based on the differences identified and pronounced within a set of human behaviour. Religious extremists devalue other humans who do not profess their beliefs, just like racial, class, and ethnic extremists.

Osama bin Laden exploited these socially constructed divisions to manage his terrorist organization, al-Qaeda, which utilized existing religious differences to commit terrorist activities. In Somalia, the war first exhibited itself as inter-clan conflict arising from the exclusion of certain clans by the ruling government in administration of the state. In Rwanda, the genocide of 1994 was based on ethnicity, just similar to any racial discriminatory atrocity. The Hutus have features distinctively different from the Tutsis and the two groups exterminated each other. In Kenya in 2007, it was ethnic violence and mostly about the language being spoken rather than the characteristics which are almost similar across.

While terrorism has caused reductions in human interactions due to restrictive activities of antiterrorism and counterterrorism efforts, there has been a gradual reduction of the generalized labelling of believers in Islam as terrorists. Although most known terrorist attacks have been perpetrated in the name of Islam, its condemnation by Muslims themselves has been useful in the reduction of the gap that these acts were intended to create—a wedge between Muslims and the rest of the world.

The modern and contemporary mention of the word *terrorism* is almost synonymous to Osama bin Laden and al-Qaeda for the 9/11 attacks on the Twin Towers, which changed the world completely. The effect of the 9/11 attacks has been felt throughout the world in relation to international relations, human rights preservation, and homeland security behaviour of all states. The result has been justification of counter violations of human rights, such as the Guantanamo Bay detention camp, the invasion of Iraq and persecution of Saddam Hussein, increased limitation on movements of people due to suspicion, massive investments in counterterrorism equipments, and increased suspicion by others on Muslim faithfuls. Today, if terrorism were to start producing negative effects on the faith by causing a reduction in the numbers of faithfuls in Islam or in any other associated organization, including some Christian sects, then these effects will trigger the end to religious-based extremism.

For the case of Eric Garner, which represents racial extremism at the moment, it is the response arising from citizens' united disapproval that is likely to change the racial relations and policies that encourage race-based discrimination. Hopefully, the case of Eric may become a trigger for consciousness to sanity and a concrete effort to end racial violence. Demonstrations against Eric's death were largely multiracial and an expression of non-indifference; all races came together to condemn discrimination and the extreme behaviour of those indoctrinated to devalue people they perceive as different from them.

CHAPTER 6

THE ARRIVAL OF *HOMO X*

Introduction

The answer to the question 'Who is *Homo x?*' is found in men and women whose characteristics have inherent psychological, physiological, and social behaviour differences with the *Homo sapiens* or Homo sapiens sapiens. However, within the final stages of this maturity period of the *Homo sapiens* evolution cycle, majority of the human population is still comprised of *Homo sapiens* as the dominant species making the distinctions slightly challenging. Our social lenses have not yet allowed us to look at the diverging difference between us and our youth and children. It is not possible to differentiate certain emerging traits in these categories out of the majority of us since we easily label them 'abnormality' and where there appears to be slight changes we console ourselves that these could be mere normal deviations that are 'insignificant'. Certainly, the ecosystem prepared over the years, as described in the previous chapter, has been so as to prepare the world environmentally and socially for the next level of human beings.

World leaders have called them the next generation; marketers have called them in infomercials as generation y, x e.t.c. The converging truth is that the generation so labelled has shown significant difference from previous generations in many ways to attract further inquiries.

In conformity to the Darwinian *organic theory of evolution*, with a divergence in methodology of the enquiry process, three major distinctive areas emerge that differentiate completely between the current generation of our youth and children which I have labelled *Homo x*, and the *Homo sapiens sapiens (we are)* as was the distinction between *Homo erectus* and *Homo sapiens* during the previous evolution periods. These three are (1) the substance of the brain, (2) changes in physical characteristics, and (3) social behaviour.

Evolution of the Substance of the Brain

After complete encephalization, the brain of the *Homo sapiens* has been estimated to be 1,300 cranial capacity which is bigger than their predecessors, who had as low as 900 cranial capaciy With this size of the brain, it was possible for humans to retain more memories and increase their intelligence, making them more capable to dominate their environment and explore other creative ways of survival. This increased the capacity to survive within the environment, which also affected humans physiologically as more had been learnt and innovated over the years, leading to the expansion of the physical matter of the brain size and the human head of the hominins from 1,100 cranial capacity to between 1,300 cranial capacity and 1,500 cranial capacity today.

Biological scientists and psychologists who have studied anatomy of the brain over the years assigned certain roles to some parts of the brain as illustrated in the table;

Anatomy of the human brain

No	Area	Function
1	cerebral cortex	• voluntary movement • language • reasoning and thought
2	Cerebellum	• balance • movement and posture
3	brain stem	• breathing • blood pressure
4	Hypothalamus	• thirst • body temperature • emotions • circadian rhythms
5	Thalamus	• sensory processing • movement
6	limbic system	• emotions
7	Hippocampus	• learning • memory
8	basal ganglia	• movement
9	Midbrain	• vision • audition • movement

However, these studies have not yielded much in the determination of the relationship between the behaviour of the brain over the years and the environment in detail. Not much has been recorded as the psychologists still hold that the internal functions of parts of the brain remain constant as tabulated above. For instance, if the cerebral cortex expanded, what changes occurred to the other parts like the hypothalamus, limbic system, etc.?

For the contemporary human, the surroundings have become much more complex, requiring certain adaptation to be able to cope with the challenges faced. Having overcome the physical challenge through development of tools and technology, what remains unexplored is the psychological component of the human brain's adaptability to cope. The mystery of the unexpanding cranial capacity and a constant unchanged average brain size, which is contrary to the theme of the organic theory of evolution of man, is what marks the distinction of the new form of evolution and character of the *Homo x* that can be distinguished from *Homo sapiens*.

The matter of how the brain is capable of coping with the current complex bombardment of experiences that are all necessary for individual survival in relation to the evolution discourse is no longer found in seeking physical expansion of the brain of the Homo Sapiens Sapiens evolution to Homo x in as it was for the previous hominines but in the psychological adjustments of the substance of the brain, which seems not to be changing in form and not shape or size.

Outsourcing Capabilities

While there have been several theories and myths about the capacity of the brain utilized by man, even scientific revelations, such as the physiology of brain mapping, suggests that all parts of the brain have a function without revealing the percentage of the brain in use at a time. But going by the organic theory of evolution, isn't it clear that the brain size is adapted to its functions? If the size of the brain has been increasing to accommodate its requirements, then it is safer to conclude that the brain is utilized by humans 100 per cent. But one may wonder why, even with the increased information, experience, and accumulated knowledge, the average size of the brain is still the same in size.

The experiment below seeks to demonstrate the changes in brain substance that now defines the new human, *Homo x*.

Illustration of the nature of evolution of the brain

Step 1

STEP 1

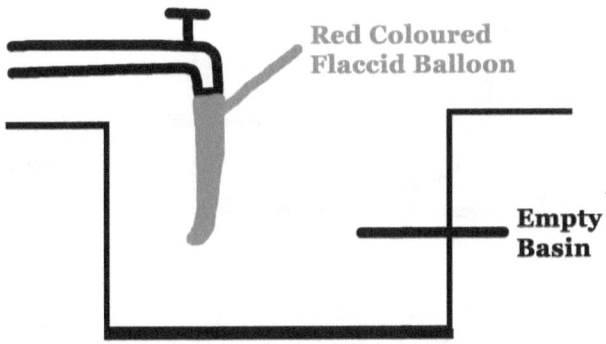

Red Coloured
Flaccid Balloon

**Empty
Basin**

1. Place a flaccid colourless balloon on a trough (both colourless).
2. Insert red soluble ink powder on the flaccid balloon (as it remains flaccid).
3. Place the trough to a water source (tap) and the balloon directly to the tap.

Step 2

STEP 2

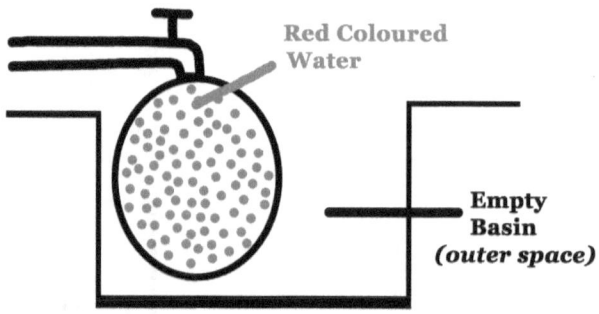

Red Coloured Water

Empty
Basin
(outer space)

4. Add water gradually to the balloon, and ensure the red ink powder fully dissolves.
5. Continue to add water until the balloon expands to its elasticity limits.
6. Stop the flow of water once it has reached its elasticity limit, and observe.

Observation: The water content in the jar is red.

Step 3

STEP 3

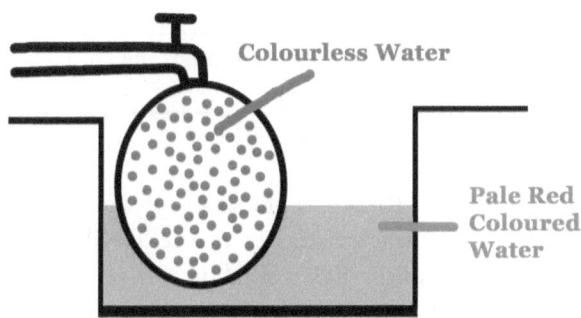

Colourless Water

Pale Red
Coloured
Water

7. Loosen your grip on the balloon to allow water to overflow as you continue to add water to the balloon.
8. Add more water to the jar until it overflows.
   Observation: After a period, the water in the balloon is colourless.
9. Stop the flow of the water from the tap.

Further observations: while the water in the balloon has turned colourless, the water in the trough is pale red.

Interpretations

Balloon:	represents the brain and how it expands over time
Red soluble ink:	represents the initial content of the brain devoid of experience or when tabula rasa, biologically defined as the anatomy of the brain

Water:	represents experience and knowledge which has expanded over the years; experience and knowledge is external to the brain but acquired
Trough:	represents the external environment where the brain operates
Tap:	represents channels of acquisition of knowledge.

Explanations

1. The organic theory explanation of the changes in hominin characteristics from the apelike creature to *Homo sapiens* has well shown that there was a relationship between the brain expansion, cranial capacity, and experiences arising from technological developments by the hominin himself. This is illustrated by the increase in the size of the balloon as more water is added into it.

2. The expansion of the brain is similar to the elasticity of the balloon. The fact that the brain has stopped expanding means that it has reached its elasticity peak, just like the balloon reaches a limit, beyond which it bursts. In relation to the experiment, to prevent the burst, the hold on the neck of the balloon is loosened to allow overflow of water to the trough. It is hypothesized that a similar behaviour has been adapted by the brain to stop its expansion and still function to accommodate its experiences.

3. The trough is the external environment or the artificial site that the brain stores knowledge and experiences, including the control of and execution of its other functions. On inventions that changed the *Homo sapiens*, I mentioned a pen and a book or the early tablets, scrolls, etc. which were used to store information.

Effects of evolution on the anatomy of the brain:

4. A look at the brain anatomy would reveal that the initial storage ability of the brain must have had initial effects on the hippocampus as humans began to store knowledge and information.

5. The rational consciousness stage of evolution, which is full of enquiry and mental exercises resulting to most of the modern innovations, increased some use of the cerebral cortex. And at some point, especially after complete discoveries, there was so many functions for the cortex that an outsourcing mechanism was necessary to prevent a brain burst. An example of an invention that supports the thought process is the calculator. The exercise of the brain in the thought and rationality process that is required to make logic of numbers can now be resolved without rigorous utilization of the rational and thought cortex. But the calculator is just a basic example of gadgets that have replaced all the prior functions of the brain. The computer is the major one. It is the source of artificial intelligence. We have developed different software that complement our thinking almost completely.

I was playing chess on the computer and I first chose to play offline on the games program that had been installed on my laptop. It has three levels—from a simple to complex game. Out of five games, I checkmated the game thrice, while I was defeated twice. Then I got online and started playing with someone in USA. I got the same result. I noticed that when you are playing against the computer program, you engage your brain just as if you were playing with another human being.

Offline, what the computer was engaging was similar to the output of a human utilizing the cerebral cortex or the thought process. But this game was prepared to respond to certain moves by someone who may not have been there but had transferred this thinking ability to the device.

6. The balloon, even after inflation, remains a rubber chemically. This is similar to some components of the brain that only adjust but remain constant in form and functionality. It is expected that there are constant highs and lows of human temperatures that support the biological functioning of other organs of the body, vision, movements of parts of the body, and the sensory glands. The basic physiological components of the hypothalamus, basal ganglia, and midbrain will remain unchanged.

7. Television sets, computers and especially the mobile phone applications that allow social media interactions have had adverse effects on the limbic system and the emotional function of the hypothalamus. This is due to reduced direct human interactions.

One day I decided to take my sons out to a children's play park in Nakuru Town. Mike is 11 years old; Cornel, 9; and Lawi, 5. The park has a bouncing castle, trains, and all sorts of children's toys.

While at the park, I asked them to have a bite first before they went to play, so we sat on the restaurant side of the park and made some orders. During the one hour spent on the dining table, my kids were busy playing games on my touch screen phone in turns. I had intended to have this time to interact with them and study them, yet they did not even pay attention to me. I thought Cornel's favourite was chicken pizza, but he did not seem to enjoy it as his eyes were glued on the phone game they were playing.

Then I decided to buy a newspaper to read and ignored them. After two hours in the park, none had gone to mingle with the other kids. But this was not an isolated case as I could see a child on the adjacent table struggling to take away her mum's phone. On our way back, I decided to just ask if they enjoyed the games at the park. They all retorted in unison, 'Daddy, the park is boring!'

During our childhood, mostly kids from privileged families could enjoy outings which we longed for. Indeed, this experience made me see how we are nurturing emotionless kids. They have little emotional connections with other people and even the external environment.

****

The trough:

8. Within the external environment, knowledge and experience has been pooled to be accessed by all making one universal pool of information and knowledge. See diagram below:

STEP 4

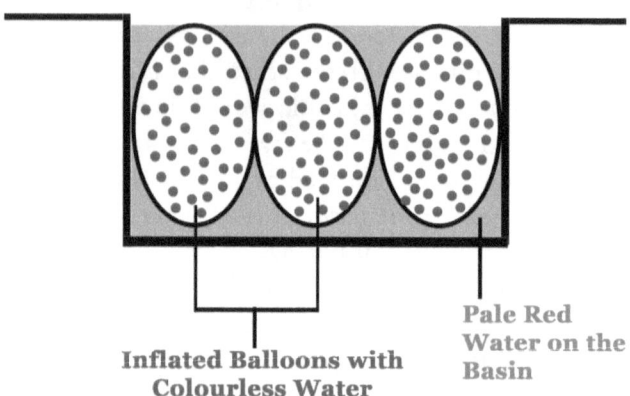

Inflated Balloons with Colourless Water

Pale Red Water on the Basin

The behaviour of outsourcing is now inherent in our kids such that at a youthful age, they are capable of sharing the same pool or extracting information and functions from a joint pool at a global level- The virtual space. This trough further makes the contemporary youth a global person, whose knowledge and experience transcends his environment and is shared universally.

9. The content or redness that constituted the initial characteristics of the fully inflated balloon and the increased water flow that still makes the water colourless indicates what has become the role of our brain—like a clearing house. This is the only way it can deal with the many experiences and challenges that require thought processes and information storage.

10. The hippocampus and any other part of the brain that was previously in charge of memory has become

a catalogue and not as a library of knowledge, as it had previously functioned.

My late grandfather, Mr Obado, had at least thirty cows. During his time, writing materials were not easily available, nor did he know how to write until late in life. For him to keep a stock record of the thirty animals, he gave each one a name just like humans, and he could identify them with their skin colour, eyes, and other features the way we distinguish one human from another. His stock record was in his brain. Just as I explained earlier, before I went for my formal education, my grandmother had narrated to me the history of my lineage up to my 20th great grand parents. Then it was possible for me to cram and put all that sfuff in my memory Forty years later after my grand fathers death, I visited a friend's ranch that had forty goats. All his goats were branded for stock records. He said he could detect loss of one of his cattle only by counting and confirming the stock numbers. He related with his cattle in terms of numbers. And that is what happens in agricultural practices today.

What does this mean? After counting and depositing the figures in an external recording material, the brain gets detached from the knowledge completely. The numbers can as well be detached until they are required again for reference. The writing material or record could have taken the role of the hippocampus and some cerebral cortex roles. What remains for the clearing house is a catalogue of records. So when one wants to know how many goats they have, the brain only needs to process information that indicates where the record that reveals this information has been stored, and this record is accessible to anyone.

This explains why during our time and up to the late 1990s, a pupil in Grade 5 or Class 4 to 5 was trained mathematical mental work and able to solve simple mathematical problems mentally. Today, this is gone. I recently asked my son Mike the answer to a simple multiplication, such as seven times eight, and I was not given an answer instantly. The boy got a calculator and was able to get the answer. There is no longer the need for kids to think as much as we did, especially when there is availability of thinking gadgets. But for them to arrive to the same answer, they must know how to use appropriate gadgets, which are now self-interpreting.

On identification of the possibility that outsourcing of the anatomic functions has been enabled by the nature of evolution of the brain, implantation devices, such as microchips, have become useful for boosting the functions of brain parts or virtually all parts. Studies on the human neural system have been able to support the discovery of devices to be used to boost neural functions in consistency with the experiment on the evolution of the brain.

Virtualism

Virtualism defines the functional environment that Homo x exist in today. It means a world controlled by internet and software which is largely functional on the basis of Information and Communication Technology. It symbolizes the advancement of humans in abstract thinking and actualization of *nothingness* to be part of reality. With virtualism as opposed to metaphysics, which is basically admission that intangible ideas have no infinite rationalization, virtual thinking and undertakings are replacing the use of matter in the new regime of human development. The world has been functionally transformed into two realities—i.e. virtual reality and material reality. Both are part of the characteristics of technology that advances human evolution today.

As humans continue to realize that effects of their activities are detrimental to the environment and a threat to their own survival, innovativeness is directed to the use of less and less materials, thus enhancing virtual actualization. Several virtual products that have been seen to reduce exploitation of materials are used for communication, storage, and artificial intelligence. All these products are in one way or another a by-product of the outcome of evolution on man and a mechanism to cope externally where the biological coping mechanism has reached the peak of its elasticity.

The use of materials for communication is diminishing. Paper, ink printers, transportation of information from one location to another (which would require the use of human resources and transport materials) is no longer necessary. Sending messages and real-time communication is now done virtually. Further, human beings have become a virtual being capable of assigning himself to be represented in several virtual persons at the same time.

On storage of information, the ability of man's artificial capacity to store information is now infinite, having moved from the initial material storage, which required more space, to a virtual space that requires nothing in terms of materials if we compare material input against information stored. Further, a global virtual storage system that shares information across the globe through Internet connectivity and storage systems on various domains in the *World Wide Web* have expanded storage capacity. Even with volumes of information, virtual search engines, such as *Google, Yahoo,* and others, have ensured ease of access to information available in the infinite virtual pool.

In governance, the broadened global space for interaction occasioned by free social media, has made it possible to monitor public opinon and at the same time making it possible, for everyone with policy opinion to attract and influence social behaviour.

It can therefore be concluded based on the most recent events such as the Arab spring and mostly the Egyptian revolution that the virtual behaviour now affects the physical realities of governance to the extent that the persuasive idea generates the virtual *ad hoc leaders* of various moments and specialties. As predicted by Marx and Engels, the mentality of the contemporary youth presupposes the existence in a stateless world prior to the imposed limitations of physical movement. It means that even traditional allegiance to state, nationalism, sovereignty, religion, local sports clubs, and local geographical heroes and heroines has diminished continually. Within virtualism, all are kings, subjects, leaders, and anything within all human abilities can be accessed by all.

Democracy continues to remain the best form of governance and majority of citizens aspire to attain its ideal form (in definition, as leadership of the people by the people), Yet now, *virtualism* offers opportunity to reach its ideal and end the fallacy of the traditional democratic practice that assumes that people's opinions in governance are expressed through surrender of collective sovereignty to an individual or a handful of individuals, as it is now possible to resolve the question of governance by the people directly (direct democracy) with new technological developments in ICT.

Convergence of culture

The increase in the size of the brain of successive hominins and the production of sounds, both vocal and instrumental, resulting to language gestures and dance are the beginnings of culture. Different languages therefore emerged due to different environmental experiences and objects producing different sounds. As each generations retained sound and meaning, so did the language get enriched and thus the beginning of coding. And as the language is retained, the brain capacity expands in order to adapt to increased pressure, especially during the childhood stage.

As each successive generation continued to accumulate knowledge in meaning of sounds, so did each language develop, resulting to improved community ties and cultural practices. At some point, each group developed their own cultures and languages due to variances in environmental experiences according to where they resided. As result of wars, human settlement, colonization, the spread of modern education and trade, from the mid ages todate,universalism grew stronger resulting to assimilation of some cultural practices by the most dominant cultural groups.

The Internet, satellite digitally transmitted television programs, and social media have given access to a pool of global experiences which individuals can choose to adopt as cultural practice. All previous cultural behaviour used to identify groups or varieties of human beings are no longer in existence.

Before the great migration by some groups of *Homo sapiens* during the early archaic stage of evolution, the hominins lived in the eastern and central regions of the African continent. There is a possibility that during that period, all *sapiens* shared a similar cultural and social environment, making them the same in behaviour. It is the geographical separation arising from the migration and the new experiences by the migrating groups that created a departure from various groups.

Now, the hominins have overcome these geographical challenges of interaction and have now become one big global family again with more social and cultural interactions. The idea that triggered the formation of the Global Village Elders is a testament to the reality of increasing non-indifference and convergence to a global cultural value that would need some sort of traditional role of village elders at the global level to sustain it. This group was comprised of respected global leaders, such as Nelson Mandela, Martti Ahtisaari, Kofi Annan, Ela Bhatt, Lakhdar Brahimi, Gro Harlem Brundtland, Fernando H. Cardoso, Jimmy Carter, Hina Jilani, Graca Machel, Mary Robinson, Desmond Tutu, and Ernesto Zedillo.

In trade, barter trade is almost over as monetary values and wealth is kept virtually. This means that as old institutions—such as the traditional banking halls, post offices, and landline telephones—start to exit or reform, man begins a different method of economic interaction that does not involve face-to-face individual interactions but virtual interactions that allow one to interact globally without limitations, such as travel visas, geographical barriers, etc. In fact, long-distance travel would now be reserved for tourism and ceremonies.

Physiology and Aesthetics

The physiological characteristics of hominins have been transforming over time—from the apelike characteristics of the bent man to the fully erected human (*erectus*) and finally the *sapiens*. These changes have given each successive hominin a different aesthetic form. The three transiential stages of the *Homo sapiens* was therefore characterized by differences in physical and facial aesthetic outlook just as the three predecessor hominins. As we have agreed, the period for these changes, quantified in terms of lifespan experiences, could have superseded the transformation of both *habilis* and *erectus*.

Through this transitional period, the average or general physiological characteristics of *Homo x* and that of *Homo sapiens*, its progenitor, remains relatively the same. It is feasible to suggest that the flora and fauna have reached on average the highest stretch of evolution that affects their aesthetic outward appearance We therefore do not expect some dwarf or gigantic humans and trees or anything new that has not been seen before. It is the level where, as evolution continues, the average physiology is not to change into any new form.

Yet there are clear signs that humans with original *H. sapiens* characteristics are diminishing day by day. Graphically, human physiological aesthetics versus evolution has been perfectly submissive to the law *of marginal diminishing returns* since the ancient historical times, falling within the *archaic* to the *rational consciousness*, and to the *maturity stage*, which begins within contemporary times. This means that with time the average physical changes that define the aesthetic characteristics of the body of human beings have remained almost constant through evolution. This is a clear departure from the popular Darwinian ideology on relationships and succession from one species with apelike characteristics to the modern *Homo sapiens*.

However, some differences have been observed on the generations that largely use computers and mobile phones to undertake social and economic activities. These youths and kids have adapted physical features that are most likely different from us as tools continue to improve. The characteristics likely to change include the average texture of the skin due to improved habitat and use of body oils and lotions across the world; body size due to the food we eat, which are mostly processed; dental size and strength as a result of processing, cooking, and softness of what is eventually consumed; the posture of the neck as preconditioned by the computers and phone use; human motion speed because we no longer run to hunt and speed has become a sport to be celebrated; and racial convergence, resulting to an outcome that has a combination of multiracial genes as humans continue to interact socially and biologically.

Convergence of Masculinity and Femininity

The long history of women's rights movement, which culminated into the Beijing conference on women and led to the Beijing declaration and platform for action,[26] coupled with the technological advancements today, are some of the social actions that have led to the change of gender roles. Additionally, the United Nations Millennium Development Goals, which is specific to the promotion of gender equality and empowerment of women, have resulted in global change in sexual relations that manifest in the divergence of behaviour and physiology of the contemporary youth.

To demonstrate this convergence, let's first begin with technological development. Most secondary roles that were socially assigned, as I have explained above, were dependent of risks and magnitude. With technological developments, these risks have been minimized and magnitudes reduced. It means therefore that both males and females of the species can now perform the same assignments on the condition that there is provision of technology. As an evolution outcome, this implies that the male human have to shed off the masculine attributes that made them perform certain roles. Indeed, unlike other animals, humans take a longer childhood, during which adaptation and adjustments to their roles are nurtured. As these reassignments were taking place in modern historical times, the nurturing process changed, thereby predetermining the growth and biological make-up of the current youth and children.

'What man can do, woman can do' is the general social roles reassignment motto.

---

[26] http://www.un.org/womenwatch/daw/beijing/pdf/BDPfA%20E.pdf retrieved 11-11-2014.

Chapter 7

The Last of the Hominin Variety

Towards Racial Convergence and a Hominine without Race

Today, there are humans of different biological varieties or races. If archaeological facts pointing to common origin of the hominid is to go by, then there is a high possibility that the early hominid were of similar physiological characteristics and had no racial, social or biological distinctions as it has been the case in the Homo Sapiens history. This means that the change of skin colour and other physical characteristics exhibited by the various races today could be the result of the conditions experienced during the *great migration* of humans from Africa to other continents, which might have taken a span of years added to the extreme difference of climatic environments that each groups have lived in for many years.

Through gene mutation arising from environmental differences or from causes explained by; Darwin's evolution theories, other biological scientists and geneticists, the *Homo sapiens* (proper) or the *sapiens sapiens* has been classified in races that resided in distinct geographical locations for the last approximately 150,000 years. The races are classified as: Negroid (predominantly in Africa), Caucasoid (predominantly in Europe), Mongoloid (predominantly in Asia); and Americas etc.

It therefore implies biologically that thousands of years before the interaction of races, each of these races only found reproductive partnerships in those of their own genetical and physical aesthetic characteristics to the extent that 'inbreeding' within the same variety, became a cultural normality. The Indians, Somalis, early Europeans communities, and even Masais in Africa practised inbreeding within kinship. In other words, dominantly there has been racial inbreeding. This inbreeding and intra-cultural cross breeding was supported by sociocultural and religious behaviour of men wherever they were and geographical distances between the races that led to inaccessibility and suspicion of each by the other of different likeness.

The result of this behaviour is illustrated in the genetical illustration below:

Assume that the possible genetical pool combination of is ABCDEF, *and gene mutation and geographical segregation over years have separated each geographical group to possess AB,CD and EF,* which has distinguished itself as for race x = gene combination AB, race y= gene combination CD, and race z = gene combination EF (with two combination of characteristics classified as races).

Then if all traits are carried forward to the next descendant, therefore, intra-racial breeding results to the following:

The following is the gene transmission table:

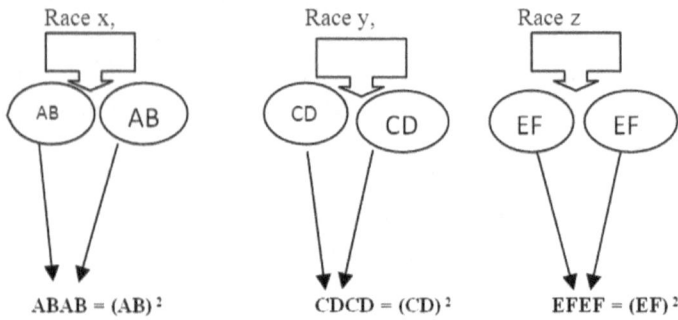

$$ABAB = (AB)^2 \qquad CDCD = (CD)^2 \qquad EFEF = (EF)^2$$

It is illustrated that over the years, determinant genes for physical features have remained the same within the various continents due to lack of multiracial interactions.

As we move forward and as explained, succession infrastructure for the next hominin (*Homo x*) has opened up a world of liberal cultural behaviour and indifference in humanity. There is no longer any cultural or racial boundary to stop or prohibit interactions of hominins.

It is these social interactions that eventually transform humans into one variety of hominins as the genetic characteristics of all hominins are combined to come up with a hybrid hominin.

Using the illustration above, we derive the next scenario to demonstrate how we arrive at a homonin without race below;

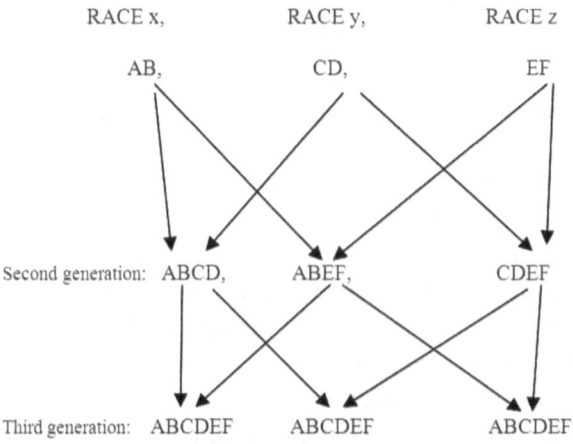

At the third level, all humans carry all traits within the possible genetical pool of hominins i.e race x, y or z are likely to have off-springs will same possible characteristics.

Physical Changes Arising from Dietary Behaviour

After my undergraduate training in 2003, I was first employed by Kenya Shell Limited as a welfare officer for expatriates and in charge of expatriate coordination for the Shell oil products for Africa in Kenya. Shell had established two pan-African offices in Kenya and in Johannesburg, South Africa, and so many expatriates were expected in Kenya.

As an expatriate coordinator, I was able to relocate and settle about thirty-two expatriates in Kenya, majority from Europe, with a few from Nigeria and West Africa. This two-year experience gave me some sense of global intercultural interaction with almost all races of the world.

The induction menu for each expatriate involved shopping assistance and a welcome dinner, in which I was deeply involved.

My observation was that the European expatriates' preference for food was more of vegetables with processed meat and not so much starchy foods. On the other hand, the African expatriates had preference for raw meat (unprocessed), more starch, and less vegetables. The Asians' diet was a combination of all and more like the European diet. But this generalization is just an observation from my experience and may not give a better picture of dietary consequences in physical similarities or differences.

There was an old lady from central Kenya (Kikuyu Tribe) whose retirement period had been extended for her exemplary work in Kenya Shell. We were fond of each other, and occasionally during work breaks, I would visit her work station for a chat. I liked her memory of history.

On one of the tea breaks, she asked me, 'Where did those tall, shiny, dark, huge, big-toothed, and deep-voiced Luos [Luo is a Nilotic tribe residing along Lake Victoria] go to?' She told me that by 1966 when she first arrived in Nairobi, it was easier to distinguish men from my community from others because of those features.

So she asked me if I was a pure Luo, and because she had never travelled to Nyanza or the Lake Victoria region, she wanted to know what had changed. This made me have a recollection of physic of my grandfather and his agemates as I saw them when I was growing up in the rural areas at the shore of Lake Victoria,It quickly came into my mind that what could have changed my generation is the dietery similarities. So I answered, 'Even the Kikuyu now eat fish,'

She informed me that she had never tested fish and that she would never eat fish. As I grew up, I learnt that every community in Kenya had their own cultural food, and as exhibited at a global level, there is Chinese cultural food, Indian cultural food,Arabic dishes,African etc. All these cuisines are now accessible to all.

Well, have you ever asked yourself whether your grandfather or great-grandfather had similar physical characteristics as you? I have. And I confirmed my former colleague's claim that my ancestors were taller than me, shiny and dark, had bigger milk teeth, and had some sort of deeper voices than we do. To confirm this further, within my family tree, I found that my great-grandfather was called Abor, which can be translated to 'I am tall', an indication that he could have been relatively taller than the average person.

Today, save for my surname, one would not be able to distinguish my cultural or ethnic background on the basis of my physical characteristics as it would be prior to our multicultural interractions.

As a baby, I fed on CERELAC like any other; our schooled parents were taught balanced diet for kids regardless of the cultural backgrounds. We had improved and manufactured foods across the enlightened masses of Africa and the rest of the world, and I was never as exposed to the scorching sun, which affected children during the days of my grandfather or my great-grandfathers upbringing with a relatively improved healthcare system.

CHAPTER 8

THE NEO-ARCHAIC STAGE

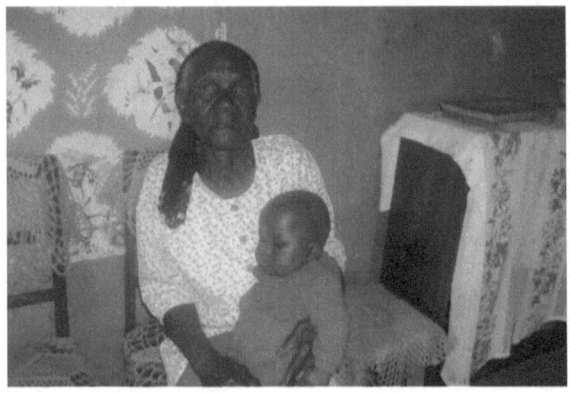

Jamal Agoya, 6 months old, on the laps of his great-grandmother, 94. The two are oblivious to the possibilities that they could be two different human species

The new archaic stage has begun. It is the period that overlaps the *sapiens'* maturity stage and the beginning of a new archaic stage of the new level of human species—*Homo x*. However, during this period, there is likelihood that

Homo sapiens is dominant in population and therefore still in control yet largely responsive technologically, socially, and culturally to the existence needs of *Homo x*.

It is for this reason that most dads and moms today find themselves sharing phones with kids or are forced to buy new phones for the kids to play games in and the youths to be glued to the phones, chatting on social media. The television sets and the programs to be watched are now dictated by these groups, Calculators and even laptops are on high demand for use by these groups of young people. Parents cannot quash these demands as they have become part of the basic life requirement for our kids.

We have moved from corpral purnishment such as caning(even though there are legislations that are prohibitive) to punishments such as confiscation of phones, denying the children any television experience, etc.?

As parents today, we experience a lot of challenges in understanding the behaviour of our kids and youths. This is because we play by the old rules of nurturing and bring up our kids with the kind of education and socialization that we had and even impose some restrictions with the hope that they will be better images of us or achieve some of our ideal aspirations.

With hindsight from the theory that our children or our youths are most likely a new human species, there must necessarily be a paradigm shift that recognizes that we are a species in transition and that the children could be of the next evolved level in the human evolution progress. Stop treating them like *Homo sapiens sapiens* because they are *not*. Your child is most likely *Homo x*, your role as a parent is now in succession management. The evolution succession management involves planning and adjusting to the new species' requirements as opposed to traditional parenting, where we guided kids and youths to fit within a set of social and environmental characteristics that was of our own creation.

153

We have to continue giving them options, both our preferences and theirs because it is known that even within the next environment of human existence, the ultimate and highest of all goals is that humans have to survive in all varieties to realize the full cycle of evolution and the outcomes of the next level. As our activities lose the real meanings because the human habitat changes, we bequeath museums of our most memorable existence of a hundred thousand years through the artefacts we have preserved over the years. Our monuments, mausoleums, celebrations, and cultural symbolisms that we observe are our hope that when learnt by the next hominins, our existence is in perpetuity.

Mr Lawrence Nyaguti (author) at the Gede Ruins in Coast Province, Kenya.

Advantages Enjoyed by *Homo x*

Firstly, the new species begins its archaic stage with an advanced maturity and with an almost perfected health infrastructure for recreation and a scientific ability to review or alter the human evolution process with technological innovations in biological research therefore making it possible for them to monitor the course of human evolution. Secondly, there are complete records across the universe of triggers, and new methods of mitigation have been found for each. The triggers for evolution in the next level may no longer be the same as man continues to conquer the earth. Triggers like wars that led to deaths and diseases will no longer affect man as there is likelihood of the employment of high-level technologies or avoidance of this risk of extermination.

The global village is interconnected with human indifference for each other, no boundaries, and accumulated limitless knowledge. Man has tested and recorded all events and outcomes for social, environmental, and biological experiments. Some of the negative outcomes have been overcome, while positive outcomes have been carried forward. The exhaustion of these ideas means none has remained unexplored, even the ones whose exploration has not been widely exposed.

In dealing with limited space for food production, genetic modification technology and cloning have outcomes resulting from genetic engineering of genes to either reproduce a copy of the same characteristics (for cloning) or, for GMO, the improvement of varieties to reproduce competitively for consumption. Genetic modification offers a variety of foods able to sustain man regardless of the season and to the quantities required even in the face of adverse climatic changes that man has no control over.

However the shortcomings highlighted in the social contradiction chapter, man has done well as a biological being.

The analogy of Evolution
The disappearance of my *childhood tree of sentimentality* into Ficus natalensis

Ficus natalensis is one of the most popularly known parasitic trees. It is either dispersed by birds on *non-self-pruning trees* or by the wind, especially when seedlings are dry. In cases where the trees have been dispersed on to a non-self-pruning tree, it begins to grow and consume food from the host tree. In such a case, it has become a parasite. The most interesting behaviour of the parasitic nature of the plant is that it grows on its host and eventually engulfs its host, leading to the death or extinction of the host. Further, the tree now remains the dominant one, standing on the previous host's position.

My father was a principal in a forestry college in Londiani, Kenya, and just at the front of our gate in Londiani, there was a beautiful tree which we used to play under. The tree, as I later learnt, belonged to the *Olea* genus, but I cannot well remember exactly which *Olea* it was. Twenty years later, when I passed by my childhood home, which was the principal's residential home, one of the features I expected to be there to rekindle my sentimental memories of my childhood years was that beautiful tree. The short shady tree was not there any more; it had been replaced by a tall, huge, and *rooty* tree. I was told it was *Ficus natalensis*, and it had engulfed and replaced the beautiful old tree.

The story of my childhood *tree of sentimentality* is the story of the constant changes in humans. I have traced my genealogy up to the twentieth generation, and every time I try to imagine the possibility of me having features or behaviours like those of my ancestors, I hit a snag. Is there a possibility that my current traits are just but a fraction of my ancestry background?

Every successive generation that progressed towards survival gradually or progressively engulfed most characteristics of its previous generations. The dominance of the next generation necessitates the disappearance of some traits—just as in the case of *Ficus natalensis* and its host. In the end, we are hosting the new generation of the new homonin that will take over from us. They will slowly but surely start exhibiting traits of their own, and once engulfed, we will disappear. We are all like my childhood tree of sentimentality and they are like Ficus natalensis in this analogy. Beneath every parasitic Ficus there are traces of its extinct host the same way humans are related to same ancestry that they are almost completely different from physiologically by genetical trace.

Exploration of the Moon and Other Planets

If anyone thought that the moon and any other space exploration to the moon, Mars, and any other planet was just human adventure, he/she is mistaken.

Evolution is basically how we transform physically, psychologically, and behaviourally to survive or as we survive. Our survival for the last millions of years on earth has helped us to transform to the level where hominins do not survive by chance but by choice.

When I shortened my discussion on metaphysical factors that may influence evolution, I meant anything out of our scope that has not yet a rational empirical explanation but is capable of physically or socially enhancing our chances for survival or even extinction. And I am certainly sure that in the event that any unknown or unperceived occurrence of collusion between the earth and any other space body devastatingly happens to the extent of loss of all lives on earth, the last creature standing will be man. The space ships would offer temporary relief for a few to begin the 'Adam and Eve' like episode. Perhaps when there comes such times when the earth is under threat by moving space bodies, atomic bombs, nuclear bombs, and any other weapon of mass destruction, they would be of positive use to mankind. Rockets and spaceships could become useful evacuation tools in case such a cosmological tragedy occurs.

The imagination of mass evacuation of humans due to the causes mentioned is not as remote. The race for acquisition of nuclear weapons by ideologically opposite states is at its highest in the 21st Century creating possibility of a nuclear war in the future that would have a devastating effect on planet earth. Unfortunately, these weapons are now possessed by states that have not been known to value human rights or some with sectorial extremist behaviour making the threat to human extinction real unless social measures are put in place to deter such ambitions.

Dear all,

RE: Call for a consensus on the binomial nomenclature of the emerging hominin

In the beginning of this book, I have referred to the next-level hominin as *Homo x*. When Charles Darwin first proposed the possibility of evolution of man, he postulated that we all come from an apelike creature. Since then, scientists anthropologists, and archaeologists have indeed proven that man is an evolving being like other animals, as demonstrated by the law of fossil succession.

By 400 BC, Heraclitus of Ephesus had come up with the doctrine of change, which has not been disapproved—change is constant. I have therefore utilized just a few aspects of recorded history to demonstrate that indeed these changes that I proclaim have occurred over the years at a magnitude that most likely we no longer understand ourselves.

What more evidence do you require? Evolution is *real*.

The first step to redefining our future therefore is acceptance of facts that we have known, and renaming this new hominin. *Homo x*, is just one of the suggestions!

A rebranded hominin necessitates the drafting of a new world order where nurturing is learning to adjust to the needs of future humans and not suffocating them with the ways that we mirror to the past. That is the only way we can sustain humanity.

It is up to us to rebrand humanity, reorient to the reality of evolution, and redesign cultural policy and any enabling infrastructure that supports an ecosystem conducive for the survival of this future hominin.

Yours sincerely,
Lawrence Nyaguti Ochieng
Human Being

References

Http://www.serpentfd.org/section2hominidevolution.html.

Bennet, John, 'Minoan civilization', *Oxford Classical Dictionary* (3rd edn), p. 985.

Hublin, J. J. (2009). 'The origin of Neandertals', *Proceedings of the National Academy of Sciences*, 106.

http://www.infoplease.com/encyclopedia/society/neanderthal-man.html.

http://www.talkorigins.org/faqs/origin.html, The Origin of Species by Means of Natural Selection or The Preservation of the Favoured Races in the Struggle for Life by Charles Darwin

archaeologyinfo.com retrieved on 24-05-2014.

Literature: Göran Burenhult: Die ersten Menschen, Weltbild Verlag, 2000. ISBN 3-8289-0741-5.

http://wafflesatnoon.com/wpcontent/uploads/2014/12/aborigines-201x300.jpg.

Abulafia, D., O. Rackham, M. Suano, *The Mediterranean in History* (Getty Publications, 1 Mar 2011), ISBN 1606060570 retrieved 26-06-2012.

Bentley, Jerry, *Old World Encounters: Cross-Cultural Contacts and Exchanges in Pre-Modern Times* (New York: Oxford University Press, 1993), 32.

http://silkroutes.net/Orient/MapsSilkRoutesTrade.htm retrieved 17- 11-2014.

Elisseeff, Vadime, *The Silk Roads: Highways of Culture and Commerce* (UNESCO Publishing / Berghahn Books, 2001), ISBN 978-92-3-103652-1.

Brantlinger, Patrick, 'Victorians and Africans: The Genealogy of the Myth of the Dark Continent', *Critical Inquiry* 12/1 (1985), 166–203.

https://www.worldwildlife.org/species/directory?direction= desc&sort=extinction_status retrieved 11-11-2004.

http://www.thefreedictionary.com/mutualism retrieved 30-10-2014.

Hobbes, Thomas, *Leviathan*, ed. Edwin Curley (Hackett Publishing, 1994).

Canada Statutes, 12 vict., c. 30: Southworth and White, 195–209; second report to the select committee on lumber trade.

http://www.cradleofforestry.com/site/home/history retrieved 11-11-2014.

http://www-naweb.iaea.org/nafa/pbg/index.html retrieved 17-11-2014.

http://www.who.int/gender/whatisgender/en/.

http://www.thebigcats.com/lion/lion_social.htm.

Martin, Hale, Stephen Edward Finn, *Masculinity and Femininity in the MMPI-2 and MMPI-A* (University of Minnesota Press, 2010) p. 310, ISBN 0-8166-2445-3 retrieved 3 June 2011.

http://www2.oakland.edu/biology/lindemann/spermfacts. htm retrieved 11-11-2014.

http://www.who.int/mediacentre/factsheets/fs241/en/ retrieved 17-11-2014.

http://www.reuters.com/article/2014/12/07/us-usa-new-york-chokehold-idUSKCN0JH2BI20141207 retrieved 9-12-2014.

http://www.odec.ca/projects/2011/yuyuya/images/ homohabilis.jpg.

http://www.bbc.co.uk/nature/life/Homo_erectus.

http://media-2.web.britannica.com/eb-media/38/79538-003-A3531D29.jpg.

https://marcivermeersch.files.wordpress.com/2012/03/ wordpress-longlin-en-in-maludong-1.jpg?w=530 retrieved 11-11-2014.

http://www.un.org/womenwatch/daw/beijing/pdf/ BDPfA%20E.pdf retrieved 11-11-2014.

http:// wonders-wonders-of-the-world-20066948-1152-864.jpg.

http://mapsnworld.com/seven-wonders-Colloseum.jpg retrieved on 03-02-2015 at 10.57 a.m.

http://www.willgoto.com/images/Size3/ South_Africa_Johannesburg_cultural_dance_festival_552 6880f20fb4465bcf3eeecd8b165be.jpg retrieved 03/02/2015 at 11.00 a.m.

http://www.reuters.com/article/2014/12/07/us-usa-new-york-chokehold-idUSKCN0JH2BI20141207   retrieved 9-12-2014.

Add:

Index